Can Science End War?

New Human Frontiers series

Harry Collins, *Are We All Scientific Experts Now?*
Everett Carl Dolman, *Can Science End War?*
Mike Hulme, *Can Science Fix Climate Change?*
Hugh Pennington, *Have Bacteria Won?*

Can Science End War?

EVERETT CARL DOLMAN

polity

The right of Everett Carl Dolman to be identified as Author of this Work has been asserted in accordance with the UK Copyright, Designs and Patents Act 1988.

First published in 2016 by Polity Press

Polity Press
65 Bridge Street
Cambridge CB2 1UR, UK

Polity Press
350 Main Street
Malden, MA 02148, USA

ISBN-13: 978-0-7456-8595-3
ISBN-13: 978-0-7456-8596-0(pb)

A catalogue record for this book is available from the British Library.

Library of Congress Cataloging-in-Publication Data

Dolman, Everett C., 1958-
    Can science end war? / Everett Carl Dolman.
        pages   cm
    Includes bibliographical references and index.
    ISBN 978-0-7456-8595-3 (hardback) – ISBN 978-0-7456-8596-0
(pbk.)   1. Military art and science–Technological innovations–Moral and ethical aspects.   2. Science–Moral and ethical aspects.   3. Military research–Moral and ethical aspects.   4. War–Moral and ethical aspects.   I. Title.
    U42.5.D65   2015
    355′.02–dc23

                                                    2015014715

Typeset in 11 on 15 pt Adobe Garamond
by Toppan Best-set Premedia Limited
Printed and bound in the United Kingdom by Clays Ltd, St Ives PLC

For further information on Polity, visit our website:
politybooks.com

CONTENTS

Preface: Be Careful What You Look For   *page* vi

1 Can Science End War?   1

2 Is War Good for Science?   26

3 Can Scientists End War?   50

4 Can Science Limit War?   77

5 What Will Tomorrow's War Look Like?   106

6 What *Will* End War?   138

Epilogue   168
Bibliography   174
Index   181

This is a book about the promise and limitations of science and scientific thinking in the context of war. It is meant to inform those conscientious scientists and engineers who are continually stymied by the political and social impediments blocking their research and development by providing a logic for external limitations for their work, and equally to those skeptical ethicists who decry the soulless march of scientific knowledge and seek refuge in a return to some mystical past. Science is the most promising route to a future of prosperity, equality, and peace, but its output will always be subject to cooption by those who see in it the surest route to political control and victory in war. We are mired in the uncomfortable impasse that we can neither dispense with science nor assign to it mastery of our destiny. What is efficient is not always effective; what is effective is not always good; and what is good is too rarely efficient or effective. Thus science alone cannot find a solution to war, nor can politics or morality by themselves.

I grew up practicing duck-and-cover drills in grade school, convinced that World War III was not only inevitable, but terrifyingly imminent. Even as a 6-year-old, I was painfully aware that hiding under my desk would be little protection from a nuclear strike, and I still occasionally have nightmares in which I am pinned under concrete and rubble while my skin sloughs away in translucent sheets.

As a precocious student, I looked to science and technology as the last, best hope for humankind. Rational minds *had* to prevail. Science was my religion, its methodology my fetish, and war my gruesome fascination.

After university, at the peak of the Cold War, I joined the US Army and was assigned to the National Security Agency as an analyst. I took computer courses on an Apple II-c, and soaked up graduate seminars in systems theory and global cultures. I did well enough that I eventually became the first Army lieutenant assigned to United States Space Command/North American Aerospace Defense at Cheyenne Mountain in Colorado. Naturally attracted to science fiction – my boyhood heroes were Asimov, Clarke, and Heinlein – I felt I had a home, and transitioned to civilian service.

When the first Gulf War kicked off in 1991, I was placed in charge of a crisis team looking for SCUD missiles. For my efforts I was rewarded with the time and resources to pursue a doctorate in international relations at the University of Pennsylvania, with a focus on world space systems and policy. The department's remarkable faculty opened my mind to further scientific study – most especially social conflict and war – but more so to the philosophical underpinnings of political issues. I recognized that I had focused too closely on the *how* of war and science, and not the *why*.

After teaching political science for several years at public and private universities, in 2001 I joined the faculty of the US Air Force's School of Advanced Air and Space Studies. This remarkable school accepts 45 of the most decorated and exceptional mid-career officers in the armed forces of the US and a dozen allied nations, and subjects them to a year of intense liberal arts education in history and the social sciences – unique in their combined experience with professional military training. In my 14 years of exposure to these magnificent men and women, I have learned far more than I have taught.

Although based on more than 30 years of study and experience with scientific and military organizations dedicated to the practice of war in the hope that it might ensure peace, this book would not have been possible without the encouragement and critical examinations of my editors at Polity, the incomparable Emma Longstaff and Jonathan Skerrett, as well as Christopher Coker and Michael Evans who reviewed the initial draft and made enormously helpful comments.

ONE

Can Science End War?

"Now I am become Death, the destroyer of worlds."
J. Robert Oppenheimer

On July 16, 1945, J. Robert Oppenheimer, the father of America's atomic bomb, was a "nervous wreck" (Morton Szasz 1992: 71). He was chain-smoking and his weight had dropped precipitously, to less than 115 pounds. General Wesley Groves, Commander of the Manhattan Project, was so worried that he had recently ordered mandatory psychological evaluations for all his top scientists to determine if they were still mentally stable enough to carry out their duties (Clarfield and Wiecek 1984: 51). It seemed as though nothing would go as planned. Preparations for the first atomic explosion had been frantically underway at the Trinity test site near Alamogordo for months, but a rare tropical air mass had moved in the previous week, transforming the usually arid New Mexico desert into a muddy quagmire. Technicians were still making repairs after a bulldozer accidentally cut the main cable from the control center to the tower that held the bomb some 20 feet above the ground. Making matters worse, that very morning a report came in from a test of the conventional-explosives atomic detonation device (the so-called "Chinese copy," a mock-up of the Trinity bomb without fissile material) that there wasn't enough explosive power to initiate a chain reaction in the plutonium (Rhodes 1986: 656–7). Trinity could very well be a dud.

Repairs were hastily completed, the rains stopped, and fingers were crossed. Bets were made among the scientists regarding the yield of the device. Incalculable intellectual effort and more than three years of the most expensive scientific project in history were about to be validated – or discredited – in the following minutes. Oppenheimer peered over the shoulder of electrician Ernest Titterton, who initiated the firing sequence. The many high-speed cameras set up to monitor the event began filming, along with seismographs, geophones, and spectrographs, but before the confirmation signal they expected on the small screen in front of them appeared, the control room lit up so brightly that neither could see it. They rushed out the door, and as the light faded they peered through smoked glass at the rising mushroom cloud. Forty seconds later, the shock wave hit.

Oppenheimer was jubilant, as were most of the scientists and technicians at Trinity. As he triumphantly circled the room he delightedly pumped the hands of everyone he could find. By the time he reached the last member of the team, physicist Kenneth Bainbridge, the technical director at Alamogordo, the joy in the room had tapered considerably. Bainbridge responded to Oppenheimer's congratulation with a muted: "Now we're all sons of bitches" (Rhodes 1986: 675). Bainbridge might have been the first to transition fully from elation to somber grimness born of a combination of exhaustion and then trepidation, but the mood quickly overtook the rest of the group. Oppenheimer, who studied philosophy to help him understand science, was aware of the change. He had discovered the Hindu religious text *Bhagavad Gita* while at Harvard, and had taken great comfort from it. Later, at Berkeley, Oppenheimer learned Sanskrit just so he could read it in its original form. As exultation transitioned into trepidation, he recalled a line from the warrior god Vishnu, instructing the worldly Prince to fulfill his duty regardless of the terrible human cost when doing so. Believing it would assuage some worry and put the test in perspective, Oppenheimer quoted from his own translation: "Now I am become Death, the destroyer of worlds" (Giovannitti and Freed 1965: 197).

Can science end war?

Can science, in its current form, bring the scourge of war finally to an end? It is perhaps the most important question of our age, not least because it was science that delivered the means to destroy all of humanity when scientists unlocked then released the awesome power of the atom. While war has always been bloody and too often ruinous, so-called advances in the weaponization of biological, chemical, and nuclear sciences have made it conceivably total.

In theory, science *can* bring war to an end. Of course, in theory, anything is possible. In practice... not so much. The enduring conundrum for science is that every solution offered seems to generate two additional problems, each requiring its own scientific remedy. Every effort to *limit* the occurrence or destruction of war through science has in time had the unintended effect of increasing both the ways and means of *making* war. When the fixed machine gun, for example, made trench warfare dominant by World War I, military planners sought a scientific solution to overcome or bypass trenches and strike deep into enemy territory – the rationale for adapting both the airplane and the automobile (as a heavily armored tank) as warfighting

4

technologies and poison gas as a chemically derived battlefield tactic. Complicating matters, the pace of change is accelerating, meaning that ever more scientists have to work harder in a growing variety of specialties just to keep up. Today, practically invisible robotic drones overfly the battlespace, monitoring persons and locations of interest and, when conditions permit, unleashing deadly missile attacks. Terrorists put toxic gasses in subways and threaten municipal water supplies with poison, extremists execute innocent victims on the Internet and use the resulting notoriety to recruit armies of sadistic killers, and states arm rebel forces with sophisticated artillery, tanks, mines, and anti-aircraft defenses, all the while maintaining plausible deniability. Science has thus allowed war to reach the primordial depths (the deep seas), physical heavens (in outer space), and virtual domains (cyberspace), with no hint of slowing down. Such a path may follow classic economic theory (keep up the supply of scientific solutions in a profitable market while increasing the demand for more scientists), but is intuitively unsettling when adapting emerging technologies to the violent destruction of war.

The Trinity detonation of the world's first atomic bomb will be referred to throughout this text as an exemplar of the disconnect between the goals of science and the ends of war as well as the long-standing if uneasy relationship scientists and warriors have had throughout history. That relationship was solidified with the success of the Manhattan Project, and set the blueprint for all major powers' weapons development and war-planning efforts since. States will use every scintilla of scientific expertise available to them as they seek out the next technology that will change the character of war, and the full force of government will be leveraged to make sure scientists' efforts are not available to potential adversaries. For their part, scientists will flock to national banners as access to massive government funds and super-secret technology are available only through them. If the cutting edge of science is where scientists want to be, then the state-sponsored research laboratory is the place to be.

The result is that science and war have become inextricably linked, and in this book I explore the nature of these links and the problems that abound. To do so I delve into philosophy to underpin my assertions, and attempt to tie these metaphysical ruminations to the practical efforts of science in support of war, and the equally unsettling if contemporarily ubiquitous efforts of war in support of science. The bottom line is that science cannot end war, for science is less an ideology than a tool. It serves those who use its methods.

Defining science and war

We all know war when we see it. It is the most brutal form of social interaction, as a rule to be avoided except as last resort, and yet common enough that it appears to be a fixture of human development. It should hardly need a strict academic definition, but for a thorough examination of the relationship between science and war, both must be identified and judiciously separated.

The authoritative *Academic Press Dictionary of Science and Technology* defines science as "The systematic observation of natural events and conditions in order to discover facts about them and to formulate laws and principles based on these facts" (Morris 1992: 1257). This widely used definition was reworded and formalized in 2007 by the UK Science Council, making it the world's first official definition of science: "the intellectual and practical activity encompassing the systematic study of the structure and behavior of the physical and natural world through observation and experiment" (Sample 2009).

Science is therefore not properly a thing; rather, it is a *way of knowing*. It presumes a systematic and rational universe that can be studied and understood through rigorous application of the scientific method. It is evidentiary, accepting as valid only data that can be observed and measured and outcomes that are repeatable. It is further understood that the *purpose* for studying nature through science is to gain control over the physical world so as to enhance humanity's security and welfare. In principle, it is objective and value-neutral, meaning that its study should not be limited by preconceived goals or required results, though it recognizes that moral and ethical considerations *can* be externally mandated – such things as limits to experiments on human subjects, for example. Nonetheless, it is widely understood that such constraints also *limit* the potential of science to discover or reveal knowledge. It is perhaps interesting that advocates of pure scientific research commonly demand that once science sets out upon a path of discovery it should not be controlled or constrained by non-scientific sensitivities just as many military practitioners often lament that war, once declared, should be left entirely for military professionals to prosecute unfettered by political interference. Both views, as will be shown, are troubling for the investigation at hand.

The early sociologist Friedrich Engels, Karl Marx's political collaborator, defined war as *organized violence carried out by social groups seeking governing legitimacy* – that is, the right to rule. Violence is a necessary component, to be sure, and by this is meant *military* or *armed force*, otherwise one could include such trivial misnomers as wars on poverty, education, or Christmas. This definition further excludes social violence running the gamut from individual spats and deadly duels, through violent team sports such as American football, to organized criminal activity, and all the way up to armed rebellions and insurrections (groups agitating for rights or benefits rather than ruling authority). Finally, true or proper war has a political component. It is not the size of the conflict measured in death or destruction, but the issues that matter here. That it is done by a military force does not make it war, as when soldiers assist in recovery from natural disaster or engage in humanitarian relief. Unless the military is participating in an attempt to impose a *political* solution through the use of force, it is not technically war.

Accordingly, true war exists at the conjunction of politics and violence, and is legitimate *only* when guided by and constrained by a political purpose. It is this last rationale, I will assert, that so clearly limits the capacity of science to solve the conundrum of war. Unless constrained by political or social purpose, science will march headlong into the unknown, finding knowledge and adapting it as it appears, even if it leads to destruction of the species.

The problem of war and the solution of science

Can science end war? Not alone, as I argue throughout, but is this even the right question? Is *war* the problem that requires a solution? Is it always bad? It has become axiomatic to state that modern science and war are conjoined, having been birthed together in the Renaissance of Europe and raised up in the Enlightenment then Industrial periods. The needs of the one have long been satisfied by the requirements of the other, a back-and-forth union that has accelerated the pace of both knowledge and destruction for the last 500 years. Today, the potential for destruction is total, and, since the end of World War II, the impetus for military involvement has been, publicly, not to *win* war but to *prevent* it – or at least *limit* death and damage when war cannot be avoided. The current scientific trend is therefore to make war *safer*, a peculiar oxymoron when considered from past perspectives, which has at the same time made war more precisely violent and incredibly expensive.

When war is identified as the problem, it is typically perceived as a problem of diagnosis. There is an underlying cause or disease that manifests itself in symptoms. Medical sciences advocate, in most cases, both treating the symptoms to relieve the patient of discomfort and also addressing the issues that brought the sickness to the body. Consequently, there are at least two overlapping considerations when applying the rigorous methodology of science to the essential problems of war. What can science do to *prevent* war, and, failing this, how can science mitigate or *limit the destruction and duration* of war? These require first an answer to the question what *causes* war?

The responses are varied and simple enough in general, if devilishly difficult in practice. There are competing perspectives on how and why science applies, and these tend to cluster around extremely long-held convictions or categories. In one of the most cited surviving treatises of the ancients, Thucydides, an Athenian, distilled the fundamental argument in his *History of the Peloponnesian War* some 2500 years ago. In it, he identified three primal motives that drive men to conflict: fear, honor, and interest. He further insisted these instinctual passions were eternal, shared by all peoples in all times. In the millennia since, adjustments to this basic formula have come and gone, but its essence remains.

Niccolò Machiavelli, the first modern sage of war, concurred, though he placed the three motivations in proper hierarchical order: security, wealth, and prestige.

Men, he insisted, desire most to keep what they have. Until this need is satisfied it is difficult for them to move to the acquisition of more – to expand their holdings and increase their possessions. And as their fear leads them to value security first and foremost, and then their greed motivates them to look outward and as their interests multiply, they begin to desire the admiration of others to justify their gains. Their pride demands respect, tribute, privilege, and veneration. In this order, for these reasons, are wars begun or avoided – first not to lose, then to gain more, and then to be recognized as having done so the right way (Machiavelli 1960).

Science, eminently rational, offers solutions to the problems *of* war and *in* war along these very lines. First, to enhance the security of people, defenses are created that make it difficult if not impossible to be harmed. This idea of technical protection makes war moot by preventing an attacker from achieving any meaningful gain or spoils. Ronald Reagan famously appealed to this motivation when he offered the Strategic Defense Initiative (dubbed "Star Wars" by an incredulous press) as a scientifically derived concept for a space-based *shield* against incoming nuclear missiles.

Opposite the promise of impenetrable *defense* is the paradoxical notion of unstoppable offense that leads to concepts of *compellence* and *deterrence*. In this reversal of logic, offensive weaponry is made superior to any defense so that the defender is compelled to choose one of two options – surrender or die – and is thus likely to submit without fighting. War does not happen if the defender chooses not to fight. Massacre does, genocide even, but not war.

By the mid-twentieth century, the concept was elevated to destruction so profound and horrific that the inevitable nuclear outcome would keep all sides from even starting a dispute. By 1962, America had put itself in a position to directly threaten the Soviet homeland with medium-range nuclear missiles based in Turkey and hundreds of long-range B-52 bombers located in the continental United States. The Soviets attempted to even the playing field by secretly placing about 100 medium-range, nuclear-capable missiles in Cuba. Despite having a much more robust space and missile program than the US, only a handful of USSR-based intercontinental-range ballistic missiles were operational. America discovered the plan when technically illegal aircraft overflights identified suspicious construction at a variety of secure sites. President Kennedy demanded the Soviets stop construction and ensure no nuclear weapons reached the Caribbean island.

That summer America and the Soviet Union stood poised on the brink of a limited nuclear war, perhaps the closest the world has ever come. Because of its clear numerical disadvantage, Soviet Premier Khrushchev was forced to back down, and in six months was retired to his dacha. The Politburo vowed never to be in a position of nuclear weakness again, and a crash program ensued. By 1967 the Soviets equaled the US in nuclear kilotons capable of detonating on the opponent's territory and surpassed the US in total number of nuclear warheads by 1978. From a small handful in 1962, the US arsenal peaked in 1967 at just over 32,000 warheads; the USSR's inventory topped out at over 45,000 warheads in 1987 (Kristensen and Norris 2013). Regrettably, it seemed, neither side had the ability to defend from such a monstrous nuclear attack, leaving both vulnerable to complete destruction by the other, a situation they decided to perpetuate through diplomacy. Making the best of the situation, this dreadful *status quo* was formalized by international treaty. The notion of Mutually Assured Destruction, with its ironically apt acronym MAD, was the publicly stated policy of Cold War superpowers that will likely be preserved as a historical artifact of the times, a scientifically derived if bizarrely rational response to an irrational potential political action.

Next, science seeks to mitigate the problems of jealousy and envy implicit in the primal motivations of wealth or interest. Greed and need cause some individuals to take from others what they cannot make or achieve on their own. Personal wants are in this way the basis of conflict, as wealth is not evenly distributed. Indeed, politics (of which war, to paraphrase von Clausewitz, is merely an extension by violent means) can be effectively described as the process of determining *who should have how much* in a given society. The distribution of goods as a political or ethical social issue is only relevant in scarcity, when demand or desire is far greater than availability. It is widely accepted that in a world of abundance, where the needs of all are easily satisfied, there should be no basis for fighting over mere things.

If science *can* provide that abundance, it may serve as an intervening means to an externally provided (political) end. The notion of peace through prosperity is an old one, but has tended to fall short in practice as the sources of wealth and power change over time. Today, for example, oil is a resource that is limited and unevenly distributed around the world: states that have oil have wealth; those that don't incur debt. If a substitute for oil could be found that provided cheap and unlimited power to any who desire it, going to war over carbon fuels would be irrational.

Such a panacea provides its own counter-motivations, of course. It might be that oil-producing states would not wish to have their lucrative resources suddenly made worthless, and would oppose such a development, but conspiracy theory is not the subject of this little book.

Still, where people are secure in their possessions, and those possessions are far in advance of their requirements (and even outstrip their desires), there remains a basic human need for social elevation; the recognition and respect of others. Scientists are hardly alone in this motivation, but have provided numerous examples of often petty disputes over primacy – the first to discover a thing – or the distribution of credit for their roles in advancing a particular subject. Some have caused lifelong enmity, and divided scientists into bickering camps, such as whether Newton or Leibniz was the first to invent calculus. Others, for example whether Edison's direct current or Tesla's alternating current plans for electrification should dominate, devolved into ruinous financial (and even physical) battles. And so it appears that even if fear and interest motivations could be overcome through scientific achievement, science may be especially ill-suited to handling the singularly human conundrum of competing for honor or prestige.

The three essential problems all revolve around the basic question: How much is enough (Enthoven and Smith 1971)? It was precisely the question of how much nuclear "overkill" was needed to guarantee the *pax atomica* of the Cold War that divided scientists in the mid-twentieth century. The destruction of a single city ought to be enough for cooler heads to prevail. But if it was fear of reprisal that deterred aggressive action, then only maximum fear could assure maximum deterrence. The so-called balance of terror required an outcome so gruesome, and so inevitable, that the possibility of nuclear war could not even be considered. Accordingly, at its peak in the 1970s, the Soviet Union and the United States had more than 60,000 nuclear warheads ready and available for use. The explosive force of these doomsday arsenals was equal to more than 10,000 pounds of dynamite for every man, woman, and child on the planet; enough to destroy life as we knew it dozens of times over. Who knows, maybe it worked? The Cold War ended relatively peacefully, and the nuclear arsenals of Russia and America are much smaller today.

And what of wealth? Is there an amount that is enough for everyone, everywhere? Probably, but when is there enough wealth to permanently satisfy an *increasing* population? Possibly never. In practice, however, it seems that conflicts between and among individuals and societies do not end when fear of destruction or loss is mitigated and material needs are satisfied. Humans are rational to a point, which is unique to each individual and indeterminate as well. Where rational calculation ceases to persuade, then passion, morality, and desire take over. Aberrant ideology or fundamentalism rushes in to fill the void, and notorious activities such as those carried out by Nigeria's Boko Haram (most horrifically against children) or Syria's Islamic State (IS, including forced conversions and mass executions as well as beheadings and immolations on YouTube) proliferate. The irrational actions of these groups are in contrast to the extraordinary kindness of a few altruistic organizations such as the Red Cross/Red Crescent and Doctors without Borders, but these rarely seem to balance the excesses of their opposites. To the extent that science *could* eliminate the irrational human element, and in the process take away all motivation for war, it could also eliminate what it *means* to be human. The good would be sacrificed with the bad. It is with this third level of motivational structure that science can readily be seen as a solution only to problems that fit its paradigm, and to the extent that war is ultimately a political and ethical problem, science alone simply cannot put an end to it.

The problem of science and the solution of war

Modern science is as much a faith, a belief system, as any religion or ideology. It has its own tenets. At its core is a conviction that the world is perceivable through the senses, that humans are capable of perceiving it in the same way, and that through the application of rigorous, empirical, testable, and repeatable observation, all the secrets of the universe can be revealed. To a scientist, these assumptions require no justification. They are obvious. But these beliefs are also modern. To the ancients, many post-modernists, and devout followers of the world's major religions, there are simply some things that *cannot* be known. There are some things that defy rational explanation. War is one of these.

Carl von Clausewitz, the preeminent philosopher of war, asserted that the character of war may change over time, but its essential nature remains constant. Indeed, much of the argument over war's persistence centers on the claim that it is somehow natural – a primitive impetus for social violence that is inseparable from what it means to be human. The social and biological sciences look to free humanity from this impulse by altering the nature of people. Eugenics, gene manipulation, drug therapy, and the like have been tested and evaluated.

Where nature cannot be augmented or altered, the conditions that lead to a natural violent reaction are mitigated. These comprise in part mass educational programs through public schooling and media, to include: government-controlled radio and television broadcasts; indoctrination and social conditioning through repetitive propaganda and behavior modification, primarily through legal requirements (many of which are positive, such as mandatory vaccinations or demonstrating competency in order to obtain a driver's or business license); medicinal intervention, such as forced quarantines or water fluoridation; and instituting socio-political reforms that transfer physical violence to alternative venues, such as the transition to electoral democracies, where political disputes are settled through debate and tallying votes, not from bullying or strong-arming the population. These are all long-term initiatives and most are still in play.

Unfortunately, long-term prospects for the Utopian goal of a world without war are subordinated to the perceived Realist shorter-term need for military responses to external threats. For this reason the technical and physical sciences have dominated military procurement and the adaptation of strategy and tactics. The general emphases have been on making war unthinkable or unwinnable, and the policy of Mutually

Assured Destruction epitomizes this view. The first is wrapped around notions of deterrence and the latter through either irresistible force or impenetrable defense. All are leveraged by weapons or weapons-support systems adapted from cutting-edge technologies. In the meantime, the biological and information sciences are gaining influence and may offer the most promise for the next iteration of both war-mitigating and war-winning breakthroughs. Enhancing the mental and physical capability of the individual combatants, the structure and motivations of social groups, and, increasingly, understanding the importance of social and information networks in the conduct of war are the primary emphases. This endeavor to make war more effective and efficient seems counter to any scientific effort to make war less prevalent, however; the two appear incompatible. But this may not be the case. Finding the means to make humans more efficient killers implies the reverse is also possible. Humans can be made *less* warlike by reversing or counteracting the process. The inherent flaw is that science cannot determine which direction people *should* go.

What is right and what is wrong is not a problem the scientific paradigm accepts. What is good or bad is specifically removed from the process of discovery that is the scientific method. Moral judgments confound the dispassionate objectivity that empirical study requires. As input, morality limits what should and even can be studied, and science asserts its prerogative to seek the essence of nature wherever an investigation leads. As output, morality is teleological. If we accept something as good, then science is put into the service of proving or making it so. If something is preconceived as bad, then science is made to find evidence to support that conclusion. Both are flawed by design. Thus scientists do not study whether war *ought* to be brought to a permanent end, though the question whether it *can* be is fair game. What science effectively discovers are means and ways to whatever ends politics decides to put them.

No end in sight – how far can science see?

And now we see the essential limitation of pure science. War may not be soluble by metrics, advanced technologies, or precise and logical calculations. Normal science makes improvements accretively, one advance at a time, even though each advance is accepted as the next faulty hypothesis to be built upon (Kuhn 1996). By gradually revealing nature through the objective rigor of its methodology, science has been spectacularly successful in exposing the inner workings of the universe, the relations between physical things, and the development of new products and techniques for humanity's use. The body of *information* gained is immense, and more will surely follow, but science cannot claim any priority of insight into *wisdom* – the *proper* use of all this knowledge. Science can tell us what is more or less effective or efficient, but not what is good or evil, not what is right or wrong. These are determined beyond the realm of science, where *to reason* means *to calculate*, where questions posed must be stated as what can or *could* be done. Morality is determined in the disciplines of philosophy and politics, in the moral realms of religion, where *to reason* means to follow a thought to its logical end, what *ought* to be and what *should* be done.

Ultimately, wars are not fought objectively, dispassionately, and without prejudice. They are and will be subject to the very human perceptions of the combatants. Although the blight of interstate violence has been condemned and denounced by pacifists, jurists, moralists, and religious and political leaders from the earliest written commentaries, the human toll and physical devastation of war have grown inexorably. The various sciences have sought to do their part, even though they have too often had the unintended effect of exacerbating the problem. This does not mean that science has no role to play in the human ethos of war, a case made throughout the rest of this text.

The next chapter comprises a description of the general connection between science and war, detailing the historical process of science accelerating the scope and scale of war in the modern era (after 1450 ACE). By the nineteenth century, states had begun to encourage scientific efforts through political and social engineering – establishing patent and copyright laws to encourage the profit-making incentive of scientists and emphasizing scientific methods in public schools and technical academies. Chapter 3 represents a switch in emphasis from the broader focus of science's war entanglement to the very human scientist's response to weapons-making and the devastation those efforts entail.

Chapter 4 brings the argument up to date, focusing on scientific technologies and methods that are currently in use for war and several that are expected to be transformational in the very near future. If science cannot end war, can it at least make it less destructive, and if so, what are the potential ramifications of those efforts? Chapter 5 consists of a discussion of some of the more frightening science and technologies that could make war more commonplace and destructive. All of these technologies are available today, and with refinements will change not only the methods for war-fighting, but the logic and impetus for war: *why* we fight and *how we should* fight. Chapter 6 provides a potential palliative for the rather negative onslaught of the previous two. The role of science and scientists in changing the social dynamic of war through political transformation and the promise of outward-focused technologies involved in the exploration and exploitation of outer space are highlighted, and in combination could at minimum help to end war as we know it.

TWO

Is War Good for Science?

*"War's insatiable demand for bigger and better ways of
killing men and the search for the silver bullet that would
guarantee total victory were the chief impetuses for the
scientific revolution that transformed the world."*

Ernest Volkman

Italian theorist Gaetano Mosca called the study of war,
with its grisly calculations of death and destruction,
suffering and loss, a dismal science. Trapped within the
passionless, objective search for gains in efficiency and
effectiveness through careful observation, research,
experimentation, and peer validation that has prompted
unprecedented growth of knowledge, wealth, and power
is the very human response to war's aftermath – grief,
heartache, and remorse.

But it wasn't always this way. Contemplative study,
whether in philosophy, military strategy, physics, or
the arcana of alchemy was a heady intellectual pursuit,
a gentleman's art. War could be rational, honorable,
gallant, even uplifting. Experience was the great teacher,

of course, but academic study could suffice as a dim tutor; studying war was a preparatory course and a contemplative palliative after the fact. None of these alone or in combination could account for true military genius, however. The infamous *coup d'oeil* of Napoleon, the ability to see at a glance the disposition of the battlespace and to know immediately what must be done required *insight* built upon study, practice, and a good deal of fortunate timing.

True innovation, discovery of some fresh new idea or technique, happened when a sudden bolt of comprehension came upon the well-trained mind academically or experientially prepared to accept it, personified by the ebullient and naked Archimedes shouting "Eureka!" as he tore through the streets of Syracuse giddy with the insight from his bath that the volume of any object could be precisely measured by submerging it in water. But, once revealed, knowledge had to be held secret, tightly controlled and passed on only to sworn acolytes and bound apprentices lest it be dispersed too widely. Knowledge gained could not be shared, peer-developed, and built upon, for it could not be owned. Any profit that came from the effort of gathering and developing it was instantly lost the moment its secret was out. It could not be objectively sought as an end in itself. Knowledge was power. It was used to control and oppress, whether through the state or the priesthood, through business and trade, or via philosophical adherence. Technical innovation gleaned in this manner could, and did, turn the fortunes of war.

Science in the service of war

War and science have been intricately linked since the earliest combatants recognized that the side with superior weapons would likely prevail, sometime in long-forgotten prehistory, and set out to gain a battlefield advantage. Birthed as twins, and growing up together, they matured in the killing fields of the ancient world. The following very briefly describes the role that science has played in enabling warfare – and in return being enabled by it – through much of history.

Technical innovation as a military and state-sponsored imperative had its most infamous early success when a modest tribe of hill people created a vast empire based on the superiority of iron weapons almost 4,000 years ago. As long as the Hittites could hold on to their secret of forging metals vastly superior to the best amalgam of their enemies (bronze), their domination was assured. Archimedes famously put his intellect to the service of the rulers of Syracuse, creating a large, adjustable mirror that could reflect and focus sunlight into a beam of light that would ignite ships in the harbor, much as a child uses a magnifying glass on an anthill (O'Connell 1989: 65). The Romans developed ballistae and the Byzantines excelled with a compound called Greek fire, a flammable mixture that could not be extinguished with water (indeed, as with a grease fire, water made the flames spread).

Innovations such as these gave their developers tremendous warfighting advantage, so long as the knowledge of their making could be restricted. Once the secret of iron-working was out, the Hittites were ruthlessly rolled back. Reviled for their harsh rule, they were eventually exterminated via the same technology they had used to gain singular prominence. Improvements that changed warfare but that were obvious upon employment – they could not be held secret once they were used – included the Greek hoplite shield (with its forearm brace allowing coordinated hoplite infantry to push the enemy with the entire weight of the phalanx formation) and the stirrup (which allowed riders to brace themselves in the saddle and thus allowed for the development of heavy or armored cavalry). Such innovation could be copied at will, and woe to the side that failed to keep up with the latest technical advances. Better – for states seeking an advantage over their enemies – was specialized and isolatable knowledge that could turn the tide of battle. Along with armies of foot soldiers and cavalrymen, bureaucrats and functionaries, throngs of alchemists and academic oddballs were paid or coerced to ply their trades in the service of technical superiority. These proto-scientists, with their esoteric laboratories and arcane wisdom intent on divining the occluded secrets of the ancients, became technical and scientific advisors to kings and princes.

And then a singular curiosity began a chain of events that cascaded toward scientific modernity. Sometime in 1248, about the time the great works of ancient Greece and Rome were returning to Europe from libraries in the Islamic world, the great English polymath Roger Bacon received as a gift a set of Chinese firecrackers (Volkman 2002: 61–2). Interest in the Far East had accelerated with Marco Polo's overland voyage, and this odd little toy inspired an intense investigation. Bacon was able to reverse-engineer the powdery substance that ignited explosively when subjected to a bit of flame, and wrote the formula in his *Epistle* on the "Secret Workings of Art and Nature, and on the Nullity of Magic" (Wigglesworth 2006: 94–5). His intellect was such that Bacon predicted all manner of weapons that would eventually come to pass, including rockets, aircraft, and cannons – all dependent upon his formula.

Thus the innovation that led to the creation of the modern nation-state, inaugurated the Industrial Revolution, and made war *total* began with a curious contraption from the mysterious East. The Gunpowder Revolution that followed would change everything, but until the explosive force of gunpowder could be contained and channeled into a specific direction, it would remain little more than a distraction in war. The solution came from a decidedly unscientific institution with a parallel need.

Life in Middle Ages Europe was dominated by the Catholic Church. The Vatican had placed itself and its hierarchy of cardinals, bishops, and priests between the lay person and God. Regular tithing and strict church attendance offered an eternity in heaven, but without clocks or widespread literacy the faithful needed to be called to worship to receive the benefits of devotion. The settled-upon technique was to ring vespers from bells mounted in church steeples, which could be made to sound the various times for special prayers, church attendance, and even public warnings (approaching armies) and announcements (the death of a king, for example). The larger the bell, the further the call could go out, but iron was still too impure in its then-current form, and so large bells cracked beyond a certain size, limiting their range and tonal qualities. Metallurgical craftsmen put their talents to the task, tripling the size and quadrupling the range of bells by the beginning of the fifteenth century. It is difficult to determine at what point someone recognized that one of these large bells turned upside down could be loaded with gunpowder and grapeshot, but the first culverin-shaped bombards used in war were undoubtedly just that (Stark 2006: 45; Landes 1983: 78–9). The formula for gunpowder had been revealed, and technology developed for an entirely unrelated purpose was eminently adaptable to war, but one more ingredient was necessary to complete the transition.

The famous battle of Agincourt (1415) had rattled all of European nobility by placing yeomen armed with the unique English longbow at King Henry V's service. As many as 18,000 Frenchmen, including more than 1,200 mounted knights, arranged themselves in preparation for a charge against the heavily outnumbered English (Volkman 2002: 38–41; Van Creveld 1991: 26). The French, confident in their numbers, advanced slowly, a recent rain having saturated the ground. At Henry's command, English longbowmen unleashed a hail of arrows that, due to the velocity and weight of the projectiles, readily punctured the nobility's armor and completely disrupted the advance. By the end of the battle several thousand French lay dead in the muck, dispatched alongside fewer than 400 English casualties. (French casualties, including prisoners of war, may have numbered up to 10,000. See Cowley and Parker 1996: 4; Delbrück 1990: 463–70.) The armor-penetrating arrows of the longbow became the staple of England's renewed attempt to seize the French crown. By 1428, England controlled all of northern France, including Paris, and much of the southwest. French King Charles VII was frantic for a technology to defeat the cursed longbow.

Desperate for victory, the French prepared for what appeared to be a relatively evenly matched battle near the village of Formigny in 1450. The English, as was their habit, put the vaunted longbow at the front of their ranks and awaited the usual French cavalry charge, but instead of doing so the French meticulously constructed two of their stumpy new bell-inspired cannons just outside the range of the English line. They tossed a mix of shrapnel (stones mostly) on top of a mysterious black powder they had poured down the gullet of the large iron jars. The English were crowded together, probably due to curiosity about the odd-looking bells, and when the cannons were finally touched off, an explosion sent gravel screaming toward the English lines. In seconds, more than 3,500 longbowmen were dead before they could set and loose a single arrow. In all, the English lost over 5,000 men. French casualties were minor, about 120, and the outcome of the Hundred Years War was forgone. Within three years the English were expelled from all territories in France. The longbow, so devastating since Agincourt, was utterly outmatched by a pair of crude gunpowder contraptions that hurled stones at high speed.

When a scientific breakthrough leads to technology that changes the manner in which war is fought, scientific reaction can have a snowball effect. Counter-technologies proliferate, and in so doing highlight more problems that science appears eminently qualified to solve. The next French King, Charles VIII, recognized the potential of these new weapons and commissioned alchemists and proto-scientists from all over Europe to develop better powder and stronger alloys. He had long-held ambitions in Italy, but had been thwarted by the massive fortifications protecting the rich cities of Milan, Venice, Florence, and Genoa. Perhaps the most powerful of these obstacles was the fortress of Monte San Giovanni, a barrier that had held the land approaches to wealthy Naples secure for over 200 years. The French cannon, with a different kind of fill, might give Charles the advantage he had so long sought.

Charles' army of experts replaced stone ammunition with giant iron balls that could smash mortar and stone battlements with ease. Finally ready, Charles led an army of 50,000 men accompanied by 36 cannons down the spine of Italy (Volkman 2002: 63–4). Having just comfortably outlasted a seven-year siege, it is likely the defenders of Monte San Giovanni were not especially worried by the appearance of the French forces, though they were undoubtedly curious about these new weapons. When the cannons were eventually ignited, the 50-pound iron projectiles accompanying massive amounts of noise, fire, and smoke blasted the mighty walls to rubble in mere hours. Naples capitulated, and the entire social structure of Europe, centered on impregnable fortresses and noble knights, was made obsolete. Feudalism ended, as the emerging way of war required peasants to arm themselves and fight for the state, not just provide labor to support the local lord's martial interests. States increased their size, modernized their economies to support large government bureaucracies overseeing the new mass armies, and by putting cannons on ships set out to bring the whole world under European control.

In this age of constant warfare, scientific advantage came from all manner of sources, and the state with the best academic minds could suddenly emerge as the next great power. The economy of gunpowder had become extremely lucrative, and the impetus of alchemy moved from transmuting base metals into gold to improving chemical compounds for combustion (making alchemy and war the forebears of modern chemistry). Within a few years it was clear the cannons had to be countered. The obvious answer was defensive cannonry, but what was the best means? Large cannons with long range could be pre-mounted on walls and parapets, eventually on ships (inevitably changing the dominant form of naval combat from direct assault via ramming or boarding to flanking maneuver for expressing maximum firepower), but this had no value in field engagements. The answer, which came from armorers in the Spanish occupation of the Low Countries, was not in ever larger cannons that were difficult to deploy and move in the battlespace, but in smaller *hand* cannons that could be carried into battle by infantrymen. Called a *harquebus* by the French, the new Spanish weapon matched the range of the longbow and, more importantly, targeted the weak point in the cannon's operation; the gun crews. Heavy enough that it required a forked rest under the barrel to hold steady, and extremely complicated to load and fire (requiring up to 90 precise movements), the hand cannon was quickly adopted and continuously updated. Within 20 years, a matchlock firing device and a trigger added to the usefulness of the weapon, and an elongated then rifled barrel added to its accuracy and range. The device still took well over a minute to load and fire a single round, but the Spanish developed tactics that allowed successive lines to alternately load and fire, thus sustaining a withering volley of shot wherever the battle moved.

In February 1525, a French force under the direct command of King Francis I laid siege to the last vestige of resistance in Lombardy, the small but strategically important fortress of Pavia, 20 miles to the south of Milan. Recognizing the criticality of the situation in Italy, the Spanish Hapsburg Emperor Charles V decided to engage the 28,000-strong French force with 23,000 imperial troops including 1,500 of the new Spanish harquebusiers (Delbrück 1990: 91–3). The battle began unexpectedly: the vaunted French cannons were not having their expected effects as the cannoneers were being cut down before they could set and fire their weapons. When Francis realized the harquebusiers were killing the crews with small hand cannons on the battlefield, he sent in the cavalry to quickly take them out. But with the French guns silenced, the harquebusiers targeted the cavalry so effectively not a single rider made it close to the Hapsburg line. With the cavalry routed, the Spanish guns turned their attention to the Swiss and German mercenaries and French infantry that made up the core of the opposing force. Within hours, the French had lost more than a third of their number killed, the majority of the rest wounded and unfit for service, and the entirety of attending French nobility – the army's leadership – had been killed or captured. Francis himself was taken prisoner and forced to sign an embarrassing treaty that gave up all French territories and claims in Italy. The Spanish and Hapsburg losses were minimal, and the balance of power in Europe shifted again.

The scientific revolution and the rise of modern liberal democracy

By the Seven Years War (1756–63), the Prussian Army under Frederick the Great had expanded beyond all previous measures. More than seven percent of the population, about a quarter of a million men, served in its military in 1760, and the armaments factories in Potsdam were producing 15,000 muskets and more than half a million pounds of gunpowder annually (Parker 1988: 148).

Field artillery had also been expanded from crude pots to mobile units with devastating range and accuracy, all due to carefully expanded and methodical metallurgical and chemical advances. The efforts of military-paid scientists were to make weapons smaller, lighter, and vastly more powerful, and these paid off handsomely in the move from fields to floorboards – to naval guns. The capacity to sail the oceans with ships filled with modern cannons revolutionized national strategy. The British fleet, the largest in Europe after its destruction of the Spanish Armada in 1588, maintained about 400 dedicated warships prior to 1790. Following more than 20 years of continuous combat in the French and Napoleonic Wars (1789–1815), Britain's fleet expanded to more than 1,000 warships by 1810 – all with steel cannons and manned by 142,000 sailors (Wawro 2000: 11).

Still, by 1815, the capacity of European states to support and maintain war had been maximized. Without significant breakthroughs to complement the gunpowder revolution, larger and more powerful forces were unachievable. Without a scientific breakthrough, a technology stalemate of sorts ensued, and warfare became rather predictable. Fortunately (or unfortunately, depending upon your point of view), two critical events happen around the end of the eighteenth century that completed the transition to modern science and war: the French *levée en masse* and the American patent system.

In 1789 the French people, mostly Parisians, rebelled and overthrew the *ancien régime*, replacing it with a radical new form of popular democracy. The French Republic, surrounded by hostile monarchies understandably concerned at the beheading of a long-standing and powerful king by a pack of revolutionary commoners, needed an equally revolutionary new form of military support. Appropriately, social science offered a rational solution. Since every adult male was now an enfranchised citizen, all adults were now responsible for the health and welfare of the state. In partial response, the National Assembly called for an immediate levy of 100,000 recruits to serve alongside the traditional regular or line army – particularly to secure the border. The combined Revolutionary Army, large as it was, was simply not effective. The massive increase of raw recruits combined with the desertion or imprisonment of most of the professional officer class meant training was rudimentary at best, nonexistent at worst.

Lazare Carnot, pressed into organizing the revolutionary military forces, insisted a pristine new art of war was unfolding. The old order was founded on complex but rote precision with drill and gunpowder weapons accomplished through brutal discipline; the new order was based on raw numbers of free citizens fighting for their own political and economic gain and thus imbued with a boundless *esprit de corps*. Carnot infamously averred that in this revolutionary new age there was no place for strategy and tactics; there were "no more manoeuvres, no more military art but fire, steel and patriotism" (Howard 1976: 80). The French Army would overwhelm its enemies with scientifically calculated metrics of supply and by the sheer power of its massed will. The transition to modern industrial society, in which *people* are reduced to interchangeable *personnel*, numbers to be crunched in business or in war, begins here.

This radical new military force was audacious and terrifying to the established monarchies of Europe. The Austrian Declaration of Pillnitz formally decreed that returning the French monarchy to its former position was in the vital interest of *all* nations. As a response, the Girondist French government declared war on Austria, further asserting its intent to spread revolution and democracy to every territory it conquered.

By the spring of 1793, France was formally at war with England, Prussia, Austria, Holland, Spain, Portugal, Sardinia, and Naples. The self-induced international crisis required a massive new increase in military power. The Regular Army and National Guard were combined into a single service, but both were still organized on the principle of voluntary enlistment. By mid-1793 the supply of spirited recruits had run out. The army was in shambles, poorly equipped, and with little or no time for training. France was being overrun on every front. The committee turned again to Carnot, its military affairs organizer, for a solution. He obliged, and proposed to the Committee a new scientific way of war. The *levée en masse* of August 23, 1793 was a call for *every* man, woman, and child to assist in the defense of France. All unmarried men from 18 to 25 were immediately summoned to combat duty, all others to military support. The French Academy of Science offered its full support to the state, and scientists were enlisted to work on problems of metallurgy, explosives, ballistics, hydrography, navigation, and the like. The first canneries, based on re-corked wine bottles in France but adapted to tins in Britain, provided troops with portable food (to reduce local area foraging and thus increase the practical size of military forces). A military research library was established at Meudon (which still operates) that devised the first military observation balloons. The *levée* was a desperate measure to be sure, but conscripts came forward and swelled the ranks. From European armies averaging fewer than 40,000 men a century before, the French Army now boasted a total strength of more than a million.

The *levée* did more than turn the French Army into a citizens' militia; previous decrees had already done that. More radically, it turned the army into a *nation* of soldiers; into a *nation at war*. The motto of the Revolution was "liberty, equality, and fraternity." What better way to rapidly and firmly instill the ideals of all three, reasoned the Committee on Public Safety, than in the cauldron of battle?

The final victory of Europe over Napoleon in 1815 reversed much of the political change that had occurred over the previous 25 years, and forced a return of the old monarchy, but the impetus toward a liberalized political structure remained. French citizens almost continuously rebelled against political centralization, most violently in 1830, 1848, and 1871 – the last date marking the final demise of the French monarchies and the beginning of permanent representative democracy. And ever since, the military power and scientific appeal of a nation in arms has not been superseded.

The French Revolution ushered in the era of *modern liberal democracy* and the era of *modern war*. The other great transformation, to an era of *modern science*, occurs about the same time on the other side of the Atlantic. The precocious United States changed the process of scientific discovery from inspiration to perspiration when it enacted the world's first comprehensive and systematic patent law in 1790.

Prior to state protection of intellectual property, an inventor had to be independently wealthy or have a wealthy patron in order to conduct research. Since the fruits of that research could be privately held only so long as no one else found out, systematic and collaborative research was rare. *Inspiration*, an abrupt association of an unexpected phenomenon with a puzzling occurrence in another area, was the source of superior knowledge, an essentially random occurrence describing the haphazard process of progress in scientific pursuits prior to the modern era. A sudden realization would overcome the philosopher, who would make an association between an observation and an idea for a new way of thinking or creating. Such a scenario describes the fanciful revelation of Isaac Newton: folklore purports he connected his burgeoning thoughts on force-at-a-distance when idly noting an apple falling from a tree. Gravity, which we presume had always existed, was suddenly *discovered*.

Inspiration was vital, because scientific research was clandestine. Should anyone discover the source of arcane knowledge, it could be used without restriction. A book could not be protected from anyone willing to print it; a song could not belong to anyone once sung.

So the final component of a new, better way of gaining knowledge was developing a new *incentive structure* for doing so.

Although monarchs had long granted monopolies and trade exceptions to various businesses, guilds, and court favorites, the first state-level patent law, intent on encouragement of the arts and sciences, appeared around 1474 when the Venetian Senate authorized exclusive profits within its territory for carefully documented inventions. Rigorous in its requirements, Venice granted more than 600 patents over the next 200 years, most notably one to Galileo Galilei in 1594 for a horse-powered irrigation pump (Biagioli 2006: 1132). The process was rigorous, as well as opaque, and only enforceable in the small but wealthy republic at the north end of the Adriatic. By the mid-1600s, several European states adopted laws formalizing monopoly and ownership rights over inventions, as did several of the American colonies. These localized and variously protected rights for exclusive profits were intended to spur innovation and creativity, but did not get comprehensive nation-state support until ratification of the US Constitution in 1789. Article I, Section 8, clause 8, specifically calls for the government "To promote the Progress of Science and useful Arts, by securing for limited Times to Authors and Inventors the exclusive

Right to their respective Writings and Discoveries." The Patent Acts of 1790 and 1793 established the means for gaining a patent and set the process within the purview of the Secretary of State (later under the Secretary of the Interior and currently under the Secretary of Commerce).

By 1836, the US had issued more than 10,000 patents (well over 6,000,000 today), and the Patent Office was redesigned to streamline and further systematize the patent process. It is no coincidence that comprehensive protection of intellectual property rights occurs first in liberal democratic states – Revolutionary France established the second national patent laws just after the Americans in 1791 – or that the intellectual revolution it coincides with preceded the more commonly known Industrial Age by several decades.

What was most radical regarding the patent laws of the US and France was not protection of intellectual property per se, but the *openness* of the process. All patents were to be available to the public, to be studied and debated, rather than secretly guarded and awarded or revoked on the prerogative of the sovereign authority. Whereas inventors and innovators had previously been *rewarded* for their efforts based on the perceived value of the idea for the benefit of the state (relative to the *costs* of offering protection for the idea), under the

American and French patent laws the idea became the intellectual *property* of the inventor, not a gift or privilege *awarded* by the state. In the fledgling democracies, individual rights including the right to property were accepted as preceding the social compact that established the rule-of-law state, and protection of individual rights is one of the primary purposes of the state.

It was not until the end of the eighteenth century that the notion that *ideas* could be property was first made *state policy* and launched the ages of industry and information. When profit can be made from the invention of ideas, it becomes possible to *create* knowledge. Before the so-called Industrial Revolution, knowledge was acquired by inspiration or revelation. Archimedes' *Eureka!* and Newton's apple were the result of fortuitous observation provoking inspiration. After 1800, it was conceivable to *consciously set out to discover or create new knowledge*, to realize a state-protected advantage in a systematically and rigorously defined *process* for discovering more ideas and inventions, and along with it the rise of industrial product development, the modern research university, and privately and publicly funded think-tanks. This revolution is inseparable from the rise of the modern scientific age. Indeed, it distinguishes the modern scientific method, and the world has been fervently accelerating its knowledge base ever since.

Modern science, liberalism, and war

Despite all of the changes wrought by the gunpowder revolution, a soldier from Charles VIII's sixteenth-century harquebusiers could still recognize the form of battle in late Napoleonic warfare, and with a few instructions could no doubt participate. The scale was larger, to be sure, and the rapidity of fire vastly improved, but the overall concepts were familiar. That all changed by the beginning of the twentieth century. Modern warfare would be disturbingly novel, and horrifyingly impersonal. The rise of modern science necessarily coincided with a rise in political liberalism, in the legal and economic equality of citizens. Science demanded competence, not birthright, and an educated population empowered for war demanded rights and privileges equal to their participation in supporting the state. In turn, science provided war-making power for the expanding nation-state.

The industrial period was thus ignited by a conjunction of military necessity, social transformation, and scientific incentivization, setting the conditions for the systematic construction of increasingly sophisticated weapons at a magnitude of scale well above previous efforts. Within a few decades, breech-loading firearms allowed for rapid-fire hand weapons and more accurate, powerful, and longer-range artillery; improved communications via telegraph allowed for real-time control of armies spanning entire national borders as military fronts; steam-powered movement of troops and supplies via railroad and ocean-going transportation were developed and deployed; all highlighted the rapid advances in scientific war making. As the state increased funding for basic scientific research, scientists themselves became integral elements in its promotion of the economic and industrial welfare of the state, initiating by World War I the blueprint for the modern military-industrial complex. How that transition occurred is better told from the perspective of those men and women who participated in it, described in the following chapter.

THREE

Can Scientists End War?

> *"Philosophy, as a science, has nothing whatever to do with what should or may be* believed, *it has to do only with what may be* known.*"*
>
> Arthur Schopenhauer

It is impossible to separate science from the scientists who practice and perpetuate it. If science is a morally neutral methodology that seeks knowledge for its own sake, without preconception of right or wrong, good or bad, what of the all too human scientists who *do* science? Can their moralities and ethical sensibilities somehow move science toward an anti-war objective? Possibly, as much as any future outcome could be planned and realized, but it seems rather unlikely. This is because the process of unrestricted science is far too intellectually alluring, too profitable, and, for the most famous scientists (the first to make a paradigm-changing discovery), too imbued with a legacy of admiration and esteem to be held back by mere moral qualms.

The best example could be the scientist that created the world's foremost institution for recognizing and rewarding peace initiatives. Alfred Nobel was a noted Swedish chemist in the mid-nineteenth century. Although scientifically trained from a young age, Nobel had a strong entrepreneurial spirit and successfully transitioned his father's small steel and explosives company into an international armaments conglomerate. Fascinated by the recent Swiss and Italian development of nitroglycerin-based explosives, Nobel was determined to incorporate the new science into his business. An immediate problem was nitroglycerin's volatile instability. Nobel was shocked by the loss of five employees – including his younger brother – in an accidental nitroglycerin explosion while testing the substance at his Stockholm facility. Determined to make a more stable compound, he invented dynamite in 1867, gelignite in 1875, and ballistite in 1887, each a more stable and powerful explosive than its predecessor (Fant 1991). Although dynamite and his later inventions became important in mining, agriculture (for clearing fields), and public works (roads, dams, and tunnels), and infamous as the anti-government means-of-choice in the then-popular international anarchist movement, the most lucrative application was always military sales.

Despite his stated belief in pacifism, Nobel founded more than 90 weapons-manufacturing plants and earned a fortune from military sales. Unexpectedly, upon his death he left a trust for prizes to be awarded annually for scientific achievement and, notably, for the person or organization that does the most to advance the cause of peace. The Nobel Prize is still the most prominent award in science, and the most sought-after in international politics – the latter despite the financial source of its very lucrative cash benefit. Perhaps Nobel was attempting to make up for the violence and destruction that flowed from his businesses and inventions, announcing an after-the-fact moral *mea culpa* for his failure to curb the march of science in the service of war. Late in his life, a newspaper ran a canned obituary of Alfred Nobel (instead of his recently deceased brother Ludwig): "Dr Alfred Nobel, who became rich by finding ways to kill more people faster than ever before, died yesterday." Nobel publicly joked about his early demise, but was apparently bothered that he would be remembered not as the great scientist but as "*le Marchand de la Mort*" (the Merchant of Death; Schultz 2013).

Recall the event that triggered the modern age. When Roger Bacon discovered the formula for gunpowder, he wrote it into his great opus, the *Epistolae*. Concerned that the formula could fall into the wrong hands, however, he ingeniously encrypted it within the text. Bacon believed that most people had a penchant to misuse knowledge for their own benefit rather than for that of others, and from this insight he pessimistically argued that eventually *morality itself would become subordinate to scientific pragmatism* (Volkman 2002: 58). He had little hope that the encryption would be sufficient to safeguard the knowledge of gunpowder in perpetuity, but for as long as possible it would remain in the hands of fellow intellectuals unlikely to become seduced by its potential.

Bacon was thus the first modern scientist that took curiosity to the logical end of investigation, and then rued his role in the effort. He was not to be the last. Leonardo da Vinci was similarly tormented by his ability to see the future of scientific warfare, and famously kept his notes in mirror-image to confuse casual readers (Mumford 1934: 85). Indeed, he was so sure that his depictions of submarines were likely to be developed and used extensively for violence that he kept much of his design secret, stating "This I do not divulge, on account of the evil nature of men" (Bainton 2008).

As the intricacies of long-distance sailing on the high seas became increasingly subject to scientific refinements, modern navies became devoted to the promise of technological breakthroughs that would provide even a slightly greater chance of survival in combat. Following the destruction of the Spanish Armada in 1588, England emerged (along with the Dutch and French) as one of Europe's premier naval powers. Intent on preserving its position, Elizabeth I beseeched John Napier, a Scottish mathematician and inventor of the logarithm (and a proto-computer machine calculator based on them called "Napier's Bones"), to put his extraordinary talent into the service of the Royal Navy (Gladstone-Millar 2006). Napier appears to have done so with zeal, creating a number of new designs for faster warships and developing an array of speculative war machines. So long as England was threatened by a resurgence of Spanish sea power, Napier worked feverishly for the state. But once he determined England to be comfortably ahead, he refused to work on war-making applications for the last years of his life (Brodie and Brodie 1973: 11). Napier was a well-known theologian and a somewhat dodgy occultist and alchemist, and his refusals were thought to be part of his desire to get on with other areas of interest. But on his death in 1617 it was discovered that his private notebooks were filled with all manner of devious designs for destructive war machines never passed forward to the state, along with his journal entries suggesting a personal fear that they might someday be discovered and released upon the world.

With global exploration by sea came discovery: new lands, new species of plants and animals, and new diseases, all contributing to a general fascination with scientific methods. In 1662 the Royal Society for Improving Natural Knowledge was founded in London. Members publicly proclaimed their interest only in *pure* science, essential research unencumbered by external influence or prejudice and dedicated solely to expanding the knowledge base of humanity (Thomson 1812). Even so, the state provided funding for the Society, and the members were all aware of the massive military build-up ongoing in rival France. The very first commissioned investigations were from the Crown for direct military support: astronomy and thermodynamics for navigation and ship building; metallurgy, aerodynamics, and ballistics for cannons; and chemistry for gunpowder.

The British Royal Society was such a boon to its military that Napoleon – who declared science "the first god of war" and opened France's premier scientific institute, the *École Polytechnique*, to educate engineers and artillery officers – determined to follow suit (Volkman 2002: 132). The Society for the Encouragement of

French Industry was established in 1800 with the mandate to support any scientific or industrial effort that directly supported the state's military. Highlighting the dual nature of virtually any scientific advance, the first invention to receive a magnificent cash award (12,000 francs, a small fortune in its day) was Nicolas Appert's food preservation system – what we call canning today (Goody 2013). Though its civilian economic advantage was clear, its military potential was particularly appealing to the emperor. Food that was boiled then sealed in an airtight jar could be nutritiously maintained for years, allowing armies on the march to travel further and much faster without relying on pre-placed and continuously replenished depots for resupply. In response, the British Ordnance Department issued a public appeal to any scientist who might be able to reverse-engineer the new French wonder-weapon, bottled food. Within a few months, Peter Durand figured out Appert's process, and then improved it by putting food not in shatterable glass, but in tin cans.

As the speed of scientific exploration accelerated, the association of scientists and war ministries kept pace. Some scientists expressed regrets at having to subordinate their primary interests to military research, but the top salaries and the best laboratories were increasingly government-supported, and if one had to dabble in war technology to get funding for pure science, so be it. Especially in new domains such as hydraulics and steam power, electric transmission of communications, aeronautics, and submersible watercraft, all of which had immediately apparent military value, industrial and commercial research had an auxiliary role.

Quite apart from large salaries and monetary awards for scientists, economic concerns had another impact on the development of new military technologies. In 1906 the all-steel *Dreadnought* battleship made its debut (Massie 1991). Armed with multiple gun turrets capable of rapidly and accurately firing armor-piercing shells up to eight miles, and the latest steam turbine engine capable of propelling the behemoth at up to 20 knots, the *Dreadnought* sparked an international arms race in which every major power in Europe and the Americas as well as Japan raced to equip their fleets with the biggest battleships. Within a year, Germany laid the keels for its first dreadnought-class (all big gun) *Nassau* battlecruiser. The US (*South Carolina*) and Japan (*Satsuma*) quickly followed suit, and within a few years the *Dreadnought* had been surpassed by *super* dreadnought-class battleships (including Britain's own *Orion*-class and the US *New York*-class battleships). The cost of building and maintaining these sea-going behemoths was immense, and several international arms race limitation agreements were proposed and signed in an effort to save the signatories from spending themselves into self-ruin.

Characteristic of the pre-World War I science efforts was the belief that the new tools of war were so powerfully devastating (or in the case of the dreadnoughts, so incredibly expensive) that no sane people would ever agree to go to war again. War between industrial states could no longer be envisioned as profitable, and going to war meant destroying one's own economy as well as the opponent's. This justification made it possible for some scientists – including John Holland, who perfected the modern submarine – to rationalize that even if their inventions had direct war-making applications, no sane leader would ever authorize them for use, and so why not do whatever is necessary to get one's passion for pure scientific research indirectly funded (Morris 1998: 177)? Of course, war did break out – and not just any war. Armed with the terrifying output of scientific laboratories, in 1914 states unleashed the awesome power of machine guns, mustard gas, breech-loading hydraulic stabilized artillery, and aircraft. The first fully industrialized and total conflict was dubbed "The War to End All Wars" – a moniker it failed to live up to.

Scientists in the Great War

World War I touched off in August of 1914 when an anarchist assassinated an Austrian archduke in Serbia. The anarchist movement had made great strides in the latter part of the nineteenth and early twentieth centuries due to Nobel's super-empowering invention dynamite. This powerful explosive allowed relatively impoverished individuals to sow violence at previously unattainable levels, and the anti-government sentiment of the anarchists transitioned to violent action throughout the increasingly socialist-leaning Southern and Eastern Europe. Outrage at the assassination caused Austria-Hungary to declare war on Serbia, followed in rapid sequence by the rest of the great powers of Europe adhering to the entangling alliances they had made over the previous years.

The war was supposed to be short. It would be over by Christmas. By the end of the year, however, it was obvious the technology of the day had outstripped tactics and strategy, and the battle-lines of the Western Front stagnated into a brutal repetition of trench-warfare attrition. States began to pull scientists from universities and corporations and place them directly in support of their militaries, and further started to identify prominent scientists or technicians that had been drafted and sent to the front lines so they could be withdrawn from the fight and put into laboratories. Scientific acumen had become fully recognized as a valuable strategic asset.

In an effort to break the stalemate on the Western Front, both sides began to develop methods for going past the trenches. The French and British led the way in the development of tanks and all participants rushed to incorporate the new powered flying machines into their plans and doctrines. But the most horrifying scientific advances caused this to be dubbed "the chemists' war" for its incorporation of the top scientific minds in developing poison gas, nitrates, and high explosives (Freemantle 2014). One of the best of these chemists was 50-year-old future Nobel laureate Walter Nernst. In the fall of 1914, Nernst had been mobilized as a courier for the German Army in Belgium, as were many motorcar owners. After the Battle of the Marne, the Germans settled into trenches that would remain relatively static for the next three years, and Nernst returned to his position as Professor of Chemistry at the University of Berlin with the express intention of devising a means to break through the enemy lines. He settled on a form of disruptive tear gas, and offered a demonstration in December of that year.

In attendance was another future Nobel Prize winner, 46-year-old Corporal Fritz Haber, attending as Director of the Kaiser Wilhelm Institute for Physical Chemistry and Electrochemistry in Berlin. Nernst was aware that gas-filled artillery shells had been banned by the Hague Convention, so he offered a canister system that simply released the gas into the atmosphere when the wind was favorable. Haber was not impressed with the tear gas, but reasoned if a deadlier formula could be employed – he preferred chlorine gas because it was heavier than air and would sink into the trenches – it might cause enough casualties to overcome the dug-in French and British. He took the idea to the General Staff and was immediately "promoted from corporal to captain, perhaps the most precipitous rise – for someone not in the royal family – in the history of the Prussian Army" (Van der Kloot 2004: 149–50). Authorized to organize a chemist's unit, Haber (Nobel Prize for Chemistry, 1918) brought on Nernst (Chemistry, 1921), James Franck and Gustav Hertz (Physics, 1925), and Otto Hahn (Chemistry, 1944). The assembled brainpower was impressive.

Preparation for the first chlorine gas attack started by March of 1915, but a series of mishaps (that killed three soldiers and incapacitated 50 more) caused repeated delays. British and French spies had seen the chemistry teams unloading and moving about a large number of unusual canisters and suspected poison gas, but since the presumption was the Germans would use artillery to deliver any gasses, and no additional reinforcements had been detected in support of a possible breakthrough, the warnings were ignored. On the afternoon of April 22, the canisters were in place and the wind was just right. The heavy chlorine gas oozed across the no-man's-land between the armies and tumbled into the enemy trenches.

The engagement went better than expected; the Allied line was broken. German troops advanced slowly after the gas had dissipated, unsure of the results. Finding no opposition, they seized two villages, more than 2,000 prisoners, and 51 artillery pieces (Van der Kloot 2004: 152–3). The Allies reported 5,000 deaths and 15,000 wounded. The German High Command had not expected such success, and ordered the troops to stop their advance as there were no reserves ready to exploit the breach. The day after the attack, the Royal Society Chemical Committee, headed by Nobel laureate Sir William Ramsay (Chemistry, 1904), was established to develop offensive and defensive counter-measures. The union of science and war under direct control of the state was from then on assured.

Poison gas became a staple of the war effort on both sides; as did the ubiquitous appearance of gas masks ever since. Once the poison gas genie had been let out, skillfully skirting the letter of international law, any pretense of following the Hague Convention dissolved. Gas-filled artillery shells became the vogue – waiting for the wind to consistently blow toward the enemy was no longer necessary – and about a third of German artillery rounds used by 1917 were gas-filled.

Alex Roland disputes the applicability of the moniker "chemists' war," but does assert that the influence of scientists in World War I was exceptional. "The Great War was the first of the total wars, in which the entire resources of the state – or very nearly so – were mobilized for military purposes. Not the least of the resources were science and technology" (1985: 262). The role of science would have been more pronounced, Roland asserts, had the military been prepared to exploit the opportunities science offered. This discrepancy would be remedied before the next Great War. Nation-states responded by establishing a number of scientific and technical institutions directly intended to force a science frame of mind on reluctant military leaders ... with relatively ambivalent results. World War I was fought, like all wars before it and (so far) all since, with the weapons and lessons learned from the *previous* war. Soldiers that had made careers in one branch, say cavalry, were not enamored enough with the new-fangled armored automobiles to scuttle their horses. It took full charges into machine-guns to question the victory of audacity and élan over steel and fire. Especially for the victors, it was difficult to make the argument in the early part of the twentieth century that the weapons and tactics that won the last war would be obsolete in the next one.

Of note, chemist James Conant (future President of Harvard University) stated that by 1915 the American Chemical Society offered its services to the government, but they were declined by the Secretary of War who said rather off-handedly that "the War Department already had *one* chemist and did not need more" (Williams 2010). This modified-Luddite view would change as a new generation of military officers entered the services in the interwar years. The American National Advisory Committee for Aeronautics (NACA) was established in 1915 under the Naval Appropriations Bill for that year, and with military and civilian members was charged with oversight of government aviation programs, all of which were military at the time (Roland 1985: 263). By 1939, every major combatant in World War I had accepted scientists' contributions and established formal organizations funneling scientific and technological research into practical solutions for military problems.

The experience of World War II, buttressed by the success of major spending initiatives, including Germany's jet engine and vengeance weapons developments and America's massive Manhattan Project, would cement the relationship between scientists, bureaucrats, and generals and set the course for their inextricable union ever since.

There were still scientists who either believed they *had* to serve military interests in order to fund their preferred civilian research goals, or were compelled to do so by the increasingly capable police states of the twentieth century. A precocious Wernher von Braun, for example, believed the only way he could fund his dream of a launch vehicle that would take humankind to the stars was to design a rocket capable of carrying an explosive payload continental distances (Neufeld 2007: 125–6). Germany had been heavily constrained in its military development by the punitive Versailles Treaty, and sought scientific specialists to come up with ideas to subvert post-war limitations. Aircraft were specifically proscribed, but von Braun's chemical rockets were not, and the Nazi government lavished funds and power on the charismatic scientist.

Von Braun had a penchant for doing what was necessary to achieve his desired ends. He joined the Nazi Party (claiming it was just for show, though photographs seem to show a very happy and dedicated SS officer) and accepted Jewish concentration camp forced labor (without complaint) to build his rockets (Neufeld 1995: 82–3). When the deadly V-2 rockets began raining down on England in late 1943, their ability to travel faster than the speed of sound meant no warning came to the unfortunate victims who might otherwise have prepared for their demise. The inability to counter these weapons combined with their sudden appearance was frightening to military and civilians alike.

The Germans used the carpet-bombing of their own cities by British and American bombers as retaliatory justification for these new terror-weapons, but they had little choice if they wanted to actually use them. Due to their horrible relative accuracy, they were employed as anti-civilian area weapons and primarily targeted the large urban areas in and around London. As the underground factories began producing thousands of V-2s, and as their accuracy improved, it seemed just a matter of time before they might turn the tide of the war. Indeed, Allied Commander General Dwight Eisenhower stated if the Germans had had an additional six months, they could have lobbed V-2s into the assembling D-Day forces and prevented the 1944 invasion of France (Pogue 1954: 134–5).

Science had become such a vital part of war – and war such a boon to scientific development – that after World War II the Americans and Russians scrambled to grab the best German science and scientists before their erstwhile allies could get to them. Von Braun was smuggled into the US where his rocket research became the basis for the American Redstone missiles that would become the foundation of both the US intercontinental ballistic missile (ICBM) force and its nascent space program. Much effort was spent cleaning up von Braun's image, insisting that he had been forced to support the Nazi government and had never held fascist sympathies. He was not prioritizing weapons development, his supporters argued, doing only the minimum required to achieve his ultimate dream – a mission to the moon. So successful was his public rehabilitation that he became a spokesman for America's space industry and a regular scientific contributor to Walt Disney's popular television series *The Wonderful World of Disney*.

Nuclear misgivings

Throughout the modern period, scientists had cause to vacillate between their objective curiosity and subjective misgivings, but nowhere has their cognitive dissonance been more on display than in the development of atomic, then nuclear, weapons. By 1940 it was clear that Germany posed a direct threat to the world. Poland had been crushed in a month; Denmark in a day; Holland and Belgium in a fortnight; and France in a stunning six weeks. Britain was managing to hold off an aerial assault – due primarily to secret radar and cryptographic capabilities – but if America did not intervene soon, the whole of Europe might fall to Hitler's blitzkrieg.

President Roosevelt responded by establishing the National Defense Research Council (NDRC). Its initial task was to identify the most promising American scientists and, first, to exempt them from selective service so as not to lose their talents on the battlefield (a gruesome lesson learned from the previous war) and, second, to assign them to appropriate military research and development projects. The NDRC organized a "massive migration of personnel to the war laboratories it set up, funding these operations through government contracts" (Williams 2010). The results were beyond expectations. Within a year, the NDRC had recruited and funded more than 6,000 physicists, chemists, mathematicians, and engineers, more than 30,000 by war's end. Killing and destruction passed all previous benchmarks and kept on going. Fearsome new weapons and awesome new applications were announced with almost clockwork reliability.

Throughout, scientists flocked to their national colors and despite numerous claims after the fact, there were *no* widespread or even notable outcries against the wholesale subordination of science to warfare, no public demonstrations or demands for peace, and no resignations by scientists who could not accept their complicity in the devastation their efforts supported (prior to the fact). In democratic states this had much to do with the authoritarian abuses of the Fascist Axis states, their demonstrated capacity for mass depravity, and the clear recognition that the war was a struggle for freedom. That would change in an atomic flash, when the thousands of scientists working on America's Manhattan Project recoiled at the horrific images from Hiroshima and Nagasaki. The exodus of science into direct war-making control would be complete by the late 1960s with recognition of the aptly named Military-Industrial Complex not only in centrally planned authoritarian states, but in even the most liberal, free market ones. The final push toward seamless scientific and military cooption and collaboration came about because of the development of the first weapon that could truly end all wars – and all of humanity in the process.

Can Scientists End War?

Around 1934, German nuclear scientists Otto Hahn, Lise Meitner, and Fritz Strassmann began following up Enrico Fermi's experiments in which he had bombarded a sample of uranium with neutrons with the intent of discovering new elements. They had numerous successes, but in 1938 Hahn and Strassmann – Meitner, a Jew, had been forced to flee to the Netherlands to avoid persecution and was then living in Sweden – performed an experiment with an unsettling outcome (Rhodes 1986: 247). Instead of resulting in a transuranic element close to the atomic weight of uranium, the bombardment produced barium, an element 36 places lighter on the periodic table. Unsure of the meaning of their results, or if the experiment had somehow been corrupted, they sent their data to Meitner, their colleague and previous co-director of the project. Meitner understood that Hahn and Strassmann had done something profoundly disturbing – they had split the atom and produced nuclear fission. Einstein's mass-energy equation was now verified in practice in the laboratory. Seeking verification, she contacted Niels Bohr, who instantly grasped the military ramifications. Bohr delivered a talk on fission a few days later, and nuclear physicists around the world took note (Rhodes 1986: 248). Einstein's predictions were correct, and if the chain reaction the Hahn–Meitner–Strassmann experiment produced could be controlled, a new super-weapon was theoretically possible. The state that achieved it first could dominate the planet.

Who would get the bomb first? Germany had the lead in theoretical physics, but not the resources to move quickly. Scientists in Britain, Russia, America, and Japan went directly to their top political decision makers with demands for funding to get the weapon that could harness and guarantee victory – for if the Nazis got there first, Hitler's vision of global dominance was more than just a dream.

A few scientists protested the military use of pure science. Albert Einstein wrote to several journals and newspapers expressing his dismay at the depths to which science had fallen in support of war making. Merely to gain money and equipment for research, scientists set aside their ethical and political misgivings and outwardly professed the neutral nature of their work. Max Planck famously quit science altogether after meeting Hitler, refusing to do any work that might support the anti-Semitic regime. Even Einstein, the world's most renowned physicist, had to flee Germany in 1932 as Nazi Party administrators began weeding academia of "Jewish science." The political purge not only hampered Germany's progress, as the bulk of top nuclear scientists in the world were German Jews, but at the same time accelerated America's efforts as many of them immigrated to the US. Still, the prestige of academic and government appointments, lucrative salaries, and the latest laboratory equipment were the lures, and many scientists rationalized they were supporting the long-term benefit to human knowledge, reluctantly accepting the short-term military advantages that might accrue to a particular government. Others enthusiastically offered their talents to the state. In September 1939, just after Germany's invasion of Poland, Werner Heisenberg presented himself to the Army Weapons Bureau and volunteered to head the team that would devise Hitler's atomic weapon (Lawrence 1998: 157). Like so many others, in peacetime Heisenberg was a scientist first, then a German; in wartime he was a patriot. Science came second.

In the United States and Britain, an influx of expatriated Jewish intellectuals and scientists were filling college departments, and most were extremely antifascist. Leó Szilárd, a Hungarian-born Jew working at the University of Berlin, left Germany for Britain the day after Hitler was elected Chancellor in 1933 (Clarfield and Wiecek 1984: 18–20). Concerned that Germany could build a nuclear bomb, Szilárd went to his friend Albert Einstein, then living in New York, to convince him to use his personal influence to get

America working on a bomb of its own. If any state could be expected to wield such a weapon with moderation, it was the liberal democratic state on the other side of the Atlantic. Szilárd showed him a letter he had penned for President Roosevelt, requesting an immediate crash program based on the capabilities of German science, and asked Einstein to sign it. He did. Einstein was extremely uncomfortable in his decision, but reasoned that the march of science was inevitable, and the only thing worse than a world with nuclear weapons was one in which only Germany had them. He and Szilárd sent a second letter a year later that finally got the US program going in 1940, Einstein stating in both letters that he "thought it my *duty* to inform the administration" of these events (Herken 2000: 229–30, emphasis added).

The Manhattan Project would eventually consume more than a quarter of the total American wartime budget, and 125,000 scientists, engineers, and support personnel. It was a capability no one else could match. The leader of the project was the German-educated American physicist J. Robert Oppenheimer. With essentially unlimited resources, solid progress was made and by early 1944 it was clear the scientists at Los Alamos were going to be able to develop a bomb, but would it be in time? Late in that year, British intelligence was able to determine that it would. The German team under Heisenberg had gone horribly off-track and was not going to be able to produce a bomb in the near future (Rhodes 1986: 402–3). The logic was now gone for the scientists who put their talents to work *for* democracy and *against* fascism. If Germany could not deploy an atomic weapon, then why should America? Despite their misgivings, only *one* scientist, Joseph Ratblat, resigned on moral grounds. Some scientists, including Szilárd, argued for an offshore demonstration bombing so the Japanese would have a chance to surrender, before actually demolishing a city with it. But most simply rationalized that they had started the job and now had to finish it, lest all their efforts be for naught. Besides, *someone* was going to develop a bomb eventually, and an alternate bogeyman, the Soviet Union, was the likely choice to do so.

When the *Enola Gay* dropped the atomic bomb on Hiroshima, three weeks after the Trinity test explosion in New Mexico verified the prototype, Oppenheimer was ecstatic. Despite the deaths of 120,000 civilians and the yet-unknown effects of latent radioactivity, Oppenheimer stood before the assembled scientists at Los Alamos and pumped his fist in triumph. The bomb worked! The war would soon be over! His excitement quickly faded as battle damage assessments came in, and the subsequent attack on Nagasaki increased the direct death toll to a quarter million human beings. Oppenheimer began to publicly and privately lament the ends to which science had so eagerly marched.

By 1946, Oppenheimer was in no mood to continue to develop more bombs, but in a meeting with Harry Truman he was confounded by the President's seeming lack of concern that the Soviet Union would be able to match America's accomplishment. Not only did Oppenheimer think it likely that the USSR would demonstrate a fission device in the near future, it was possible that it might take the next step, from an atomic fission weapon to the vastly more powerful nuclear fusion or hydrogen bomb. One of the Manhattan Project scientists, Edward Teller, had worked out the equations and theoretical considerations such that Oppenheimer already felt the follow-on system was realistically achievable. Frustrated by a perceived lack of efficacy, Oppenheimer became an outspoken supporter of nuclear arms limitations, and gave up applied physics altogether, but there were plenty of competent scientists willing to take up the call and continue developing weapons for war.

Curiously, in 1944, Hahn alone received the Nobel Prize in Chemistry for his part in the Hahn–Meitner–Strassmann experiment. He publicly claimed his singling out was unfair to his colleagues. But he also claimed he was unaware he had won the lucrative prize until after the war, and that he was astonished to hear it was for the experiment that led to the development of the atomic bomb. Filled with remorse, for the rest of his life Hahn became an outspoken opponent of nuclear weapons, reliving and repeating the angst of scientists whose yearning for objective knowledge was found after the fact to be at odds with their ethical views of military necessity.

If neither science nor scientists have been able to put an end to war, perhaps the problem is simply too acute for an absolute solution. Should science reject such an ambition and concentrate instead on limiting war, gradually reducing its impact in the hope that in this way it might eventually fade away?

FOUR

Can Science Limit War?

"Kind-hearted people might of course think there was some ingenious way to disarm or defeat an enemy without too much bloodshed, and might imagine that is the true goal of the art of war. Pleasant as it sounds, it is a fallacy that must be exposed: war is such a dangerous business that the mistakes which come from kindness are the very worst."

Carl von Clausewitz, *On War*

War is a brutal thing. Near-universal consensus is that the number of wars, their duration, and devastation has been far more than absolutely necessary. It is also widely perceived that the totality of war violence has been growing. Can science at least reverse these trends and minimize rather than maximize the effects of war? And if so, should it? Or will such meddling simply and unavoidably send war into unknown directions and previously unimagined forms of violence?

In this chapter, two avenues of investigation are highlighted: Reducing the level of violence inherent in individuals by following a medical analogy and, complementarily, reducing the destructiveness and lethality of the weapons of war.

Can medical science offer a "cure" for war?

What if you could take a pill that would not only remove all fears and anxieties but completely remove any aggressive or just unsympathetic impulses? What if the formula in this pill could be cheaply mass-produced and inserted into the water supply of every person on earth – much like fluoride in American municipalities? Would universal chemical attachment to a substance that made *any* form of aggression abhorrent, even mild unfriendliness or indifference to others, solve the problem of war?

Scientists generally accept the premise that human nature, consistent with the rest of nature, is subject to regularities and physical laws that can be discerned and understood through careful observation. They are also likely, when investigating phenomena of interest, to adapt successful methods and techniques from other scientific disciplines. One of the sciences most benefited from the cruelties of war, if beneficial is the appropriate term, has been medicine. The concentrated mass of casualties from battle gives surgical experience to doctors and the desperate need for saving combatants' lives and returning them to the front lines opens up new realms of experimentation and treatment. The role of antiseptics and anesthesia in surgery, palliatives and cures for diseases, and today enormous strides in the development of prosthetics are all traceable to wartime discoveries and practice.

The standard scientific method of medical investigation traces its history to Hippocrates, who established two primary branches of medicine called *semeiology*, the description and classification of symptoms, and *prognosis*, the art of making predictions based on an accurate understanding of the effects of those symptoms. The first allows the doctor to determine the patient's affliction, the latter allows for the likely outcome (if untreated) to be predicted with reasonable accuracy. The prognosis could then be altered depending on trial-and-error application of curatives.

Thucydides adopted the prevailing Greek view and his *History of the Peloponnesian War* (1954) can be seen as a semeiological case study of what is today called hegemonic war (Cochrane 1929: 177; Morley 2013: 68). More than simply a corrective or limited war for the spoils of victory, hegemonic war is a clash between the greatest powers for leadership and domination of the entire international system. Thucydides' *History*, the author hoped, would be used in the future as the basis for prognoses by statespersons faced with the need to administer nations caught in the fever of imperialism. To be sure, fever seems an apt social symptom of war, and is not an unusual metaphor for such violent political activity. Crane Brinton's *Anatomy of a Revolution* (1965), in which he directly equates the progress of revolution to that of a disease, is a classic example.

Classification and prognosis are helpful for understanding the likely outcomes of a carefully scrutinized medical condition or violent socio-political event, but are not particularly valuable in determining the *cause* of the disease/condition or safely finding better treatments. Trial and error has mostly errors. Today's medical professionals therefore look to do more than treat symptoms; their most effective work is in *preventive* medicine. A doctor may relieve a patient's pain with analgesics, for example, but the pain will continue unless the underlying *causes* for it are determined and the patient changes the behavior that brought it about.

If medical science can offer a metaphor or allegory to assist in the abolition of war, then the essential *causes* of war must be determined to effectively prevent an outbreak of it. The first assessment must determine if the cause is inherent to humans, and thus inextricable, or if it is external to them and therefore treatable. Some ailments are genetic, for example; a person is simply born with a predisposition towards them. Others are the result of exposure – to things or people. Often the cause is a mixture of both. When Europeans first came to the New World, for example, they brought with them diseases that had long been tamed by their collective immune systems but were unknown to the indigenous peoples. Perhaps 95 percent of Native Americans died in the first generations exposed to smallpox, influenza, bubonic plague, chicken pox, pneumonic plague, cholera, diphtheria, measles, scarlet fever, typhus, tuberculosis, and whooping cough (Dobyns 1983). The cause of the near-genocide was internal or individual predisposition combined with external or collective exposure to social factors. If the analogy holds as we extend it out, and war is like a disease, the capacity of science to cure it hinges on this crucial distinction: to what extent is it endemic to the human condition and how much is external to it?

War is widely perceived as an aggregate act of aggression, and the first investigations tend to be into the locus of that aggression. There are two implicit questions here. The first is, what causes aggression in individuals?; the second is, does individual aggression cause war? If yes to the latter, then the proper treatment for war is in reducing, channeling, or eliminating human aggression. Sub-questions then proliferate. To what extent are the decision-making procedures of states tied to the emotions of policy makers? Is there an analogy for inoculation against war, and if so who should be immunized? In a society where the vast majority are non-violent, is social violence an accident? An aberration? An unavoidable deviation by confused, defective, or damaged members?

Violence among people unfortunately appears universal; at least, every society has its share of malcontents, agitators, and other troublemakers that can be traced to the biological and psychological makeup of specific individuals, in turn producing various resentments, violent crimes, rebellions, and, occasionally, revolution. These are often enough attributable to emotional outbursts and irrational excesses. International war, on the other hand, is a highly institutionalized and rationalized extension of political action. Connecting the two is not necessarily straightforward, but generally begins with psycho- and sociological assessments of key individuals or cases. Still, such explanations for aggression in individuals have not yet proven sufficient to explain the wide variety of social violence exhibited in society, from hooliganism at soccer matches to social media bullying, sexual assault, and fist-fights. As all of these events appear in the context of specific socio-cultural political structures, it is reasonable to believe they can only be understood from a broader perspective.

By the time we get to the level of war, the highest and most complex form of social violence and the concern of this assessment, the contributing factors from multiple individual aggressions are amplified. And yet it has simply been impossible to provide a universal causal factor for war from one or even several individual character traits. In part this is because, in modern warfare, those responsible for the decision to go to war are rarely the ones who fight it. These decision makers are allowed a degree of dispassionate oversight that is all but absent in the battlespace. And the individuals who apply their aggression, their emotions stretched to the limit in combat, have extremely restricted input into the decisions to go to, and then how to conduct, war. Indeed, some forms of government can and have led their nations into war with little or no popular support.

Clearly human beings are animals, and all animals appear to have basic needs for air, water, food, protection from the elements, sleep, essential health care, sanitation, physical security, and passing on skills and training to the next generation. As the demand for these items usually outstrips supply, these needs are satisfied by many species aggressively. Some scientists (such as Bert Hölldobler and E. O. Wilson 1994) attempt to connect the instinctual roots of animal behavior to social interaction to arrive at explanations for aggressive social behavior that are genetically rather than culturally rooted. Nonetheless, the interactions between basic animal needs and what quickly emerges as the basic social needs of humans, to include a sense of group belonging or kinship, prestige, and self-actualization, are increasingly sophisticated and ultimately rational in character. On one side are scientists arguing for the inevitability in society of aggressive behavior based on animal instinct, and on the other side traditionalists arguing that in human social groups cooperative behavior is a better survival strategy than aggression. Who is right? We don't know. Whether individual aggression based on relatively instinctual motivators such as sexual appetites (causing aggressive competition that increases the probabilities for passing on the best genetic traits, for example) is the ultimate cause of war or whether war is caused by members of social organizations consciously *choosing* to fight for specifically articulated reasons in historically unique contexts is most certainly in dispute.

If war is part of the human condition, a normal outgrowth of the uncontrollably pathological, obsessively disruptive, and clinically dysfunctional nature of humanity, then the scientific solution to war is to *change the nature of people*. Eugenics – consciously weeding out aggressiveness through sterilization and forced breeding for specific traits – is one scientifically derived method. Chemical enhancements and gene manipulation are two others. But these seem quite radical to the majority who rightfully fear that such meddling in essential humanness will have inevitable, unanticipated, and completely horrifying results. If war is a human activity, one way to end it is to eliminate humanity, a decidedly unsatisfying solution but perfectly logical if war is equivalent to a potentially fatal disease that needs to be treated and ultimately eradicated or the patient will surely die. But is war so assuredly terminal? Is there no benefit to war? Are some wars *worth* fighting?

If so, then short of eliminating humanity, science cannot eliminate war. But can it at least *limit* war?

Can science make war safe?

After 1945, according to Bernard Brodie, military emphasis shifted from winning wars to preventing them

(1946: 14–16). Nuclear combat was simply unthinkable, and so efforts to limit *all* war seemed a reasonable path as any conflict could escalate to an exchange of ultimate weapons. This was most certainly an abrupt about-face, and has led to one of the great military oxymorons of all time: *can war be made safe?* The question is posed with both micro and macro intent. At the micro level, can individual casualties and collateral damage be kept to a minimum, and at the macro, can conflict be contained before it becomes a large-scale war? Since science only accepts observable, repeatable, and measurable data for evidence, the sliding scale from peaceful conflict – another oxymoron – to escalating conflict can be charted, coded, and subjected to quantitative analysis. Fewer dollars in damage and fewer and less severe casualties become the measuring stick for better and worse war.

Unfortunately, perhaps, such a view obscures the importance and meaning of war and brings up a set of ethical questions that challenge the basic premise. Is a less dangerous war really better? Shouldn't war be a frightening, horrible experience, so that we avoid it rather than make it commodious? Is safer war easier to enter into, and does it therefore make war more likely?

This approach has some serious deficiencies, or at least not such clear-cut advantages. One of the problems for state-level decision makers is a poverty of mid-range options; between doing very little and going to full-scale war, there is too often a paucity of choices. Nuclear war can be deterred by threatening nuclear Armageddon if attacked, but does the same nuclear threat work against terrorists or gangs killing innocent people in foreign lands? Does the threat of a direct, albeit precise, military strike promise to reduce extant violence or does it increase the likelihood of violence in an upwardly spiraling reaction to the strike? It is for this reason, to show that the intent to harm is limited and proportionate, and in most cases reversible, that decision makers want science to provide non-lethal weapons – *credible* coercive tools that get results but do not cause additional or long-standing animosities.

Non-lethal, reversible-effect weapons have taken time to get accepted by the defense community for a number of reasons. The end of the Cold War took the emphasis off nuclear combat and highlighted the less-than-world-war conflicts all around the globe. The infrastructure dedicated to nuclear weapons development went into a tailspin for careerists in participating national laboratories, and they frantically searched for replacement technologies. Developing non-lethal weapons was thus a bottom-up technology *push* for new policy and strategy (as opposed to technology development *driven* by strategy and policy), and so has taken time for the military's top brass to accept it.

Moreover, despite the appeal of non-lethal weapons to police organizations for crowd control and dispersion, many of the more readily acquired technologies have been eschewed by military traditionalists who find themselves heavily involved in policing chores abroad (e.g. nation-building, humanitarian relief). This ambivalence is in part due to the increasing militarization of domestic police (acquiring traditional military equipment such as armored personnel carriers, body armor, sniper rifles, and special operations gear) and routine use of the military for non-kinetic peace-keeping and political restructuring duties that has blurred the distinction between domestic police power and international military force. These developments have produced serious fissures between those who call for military effectiveness without long-term damage and the traditionalist view of war as a messy business in which people die and things get blown up. For the latter view, the purpose of military power must include the intent to *maximize* violence within the constraints placed upon it by legitimate governing authority and context, while the purpose of police power is to *minimize* violence. The techniques are usually mutually incompatible.

Regardless, as time hurries on, policy makers are sending the old military thinking of peace through the prospect of massive retaliatory violence on its way to history's dustbin. Scientifically derived non-lethal (or less-lethal or less-than-lethal) technologies appear to be going forward as proper warfighting options for the twenty-first century. Indeed, most of the positive reaction to non-lethal weapons in America's military is in support of its many humanitarian and nation-building requirements (sometimes referred to as Operations Other Than War or OOTW). Like them or not, the military tends to embrace those missions that get the most funding. The development vector is for tactical support of long-term peace-keeping and humanitarian operations as opposed to clashes of conventional battlefield forces. As the military is increasingly fielded in dangerous regions for these kinds of situations, it does begin to resemble an international police force. At any rate, it seems dubious to put trained killers and demolitions experts into a situation where they are bringing relief to the oppressed – unless, so the argument goes, they are armed with non-lethal weaponry. Of course, knowing that the opposing military is using equipment that is painful (to be sure) but not deadly or even permanently disabling, wouldn't those intent on disrupting the peace or the humanitarian effort be emboldened to do so? As briefly discussed in the preceding section on limiting war by altering what it means to be human, it is a very thin and murky line we cross when we ask science to change the very nature of war.

Categories of non-lethal warfighting weapons

The US Department of Defense (DOD) defines non-lethal weapons as those purposely designed and primarily employed to incapacitate personnel or material, while minimizing fatalities, permanent injury to personnel, and undesired damage to property and the environment. Minimizing is the operational term here, as no guarantee that persons or property will not be permanently maimed or destroyed is made. As is shown below, some of the most benign anti-personnel techniques are dangerous and even fatal under some conditions.

Non-lethal weapons technology development has been ongoing for some time, but has only recently become a major military pursuit. In such situations, when the technology is advancing so rapidly, it is useful to make general categories of development rather than try to select specific technologies that may or may not be influential in a few years or even months. With that in mind, a reasonable schema is to begin with a distinction between weapons designed to disable or disarm people and those designed to have effects on equipment and materiel (Moreno 2012). Within each category are physical, chemical, biological, and directed-energy components.

Anti-personnel *physical* weapons include rubber and plastic bullets as well as beanbag rounds from modified shotguns, foam batons (instead of wood or steel), entangling nets, and water cannons. *Chemical* weapons include tear gas (CS or pepper spray), sticky foam that expands to make movement difficult if not impossible for those trapped, neural inhibitors, sleep-inducers, irritants, various calming agents in aerosol form, and stink bombs (olfactory weapons). There are serious impediments to using chemical weapons, mostly legal interpretations. Indeed, there are no legal non-lethal *biological* weapons, but one can envision a non-law-abiding state, organization, or individual developing variants of mild to debilitating diseases that would infect the opponent. The last category, *directed energy*, is the most promising and potentially game-changing of the bunch. These include microwave beams that make an individual feel as though the skin is on fire, flash bang and stun grenades, lasers that blind the opponent (temporarily, if possible), and auditory weapons that cause pain, disrupt sleep, and slow communications.

Reducing collateral effects and destruction of property is also important. Stealth technology is the most developed here, stealth being a way of fighting that allows the attacker to get as close as possible to the opponent without being detected. As a rule, the closer one gets without notice, the more precise (and deadly) one can be in hitting the target while at the same time reducing or eliminating collateral damage. Such anti-materiel weapons fall into the same broad categories as anti-personnel weapons. *Physical* weapons include netting, made of fibers or wires that entangle vehicles and equipment so that they are unusable until a significant effort to remove them is made, particles that can induce short circuits in electrical or electronic equipment, tire spikes, and various instant set-up but completely removable temporary obstacles. *Chemical* weapons include sticky and vision/sensor-obscuring foams, friction reducers that affect brakes and maneuverability, embrittlement agents applied as a spray that are able to alter the molecular structure of metals making them weak and susceptible to structural failure, combustion modifiers in the air that stall engines, metal fibers that get into and jam machinery, chemicals that clog filters, super-corrosive materials that eat away metal, additives that cause fuel to gel or solidify, and super adhesives that temporarily stop machinery. *Biological* weapons limit damage by degrading or making unusable critical military equipment. This can be done, for example, by releasing a cloud of microorganisms over the opponent's territory that subsists on (eats) petroleum, explosive compounds, and natural rubber. *Directed-energy* weapons are also the most developed for damage mitigation, and include high-power microwaves capable of detonating bombs and mines from a distance, pulsed and directed-energy beams that disrupt radar and communications, lasers that can burn through metal components selectively, and infra- and ultrasound generators. Low-energy lasers could be valuable for disrupting visual augmentation devices, such as night-vision glasses, range-finders, and other target acquisition equipment. All of these capabilities are available today and are either fielded or in prototype testing. Just because a weapon is non-lethal does not mean it isn't extremely dangerous or alarming, however. The following section describes a number of these weapons, their intended uses, and their potential undesirable effects.

Some possibly frightening examples of non-lethal weapons

Sight weapons

Sight weapons are intended to limit the ability of the enemy to see and evaluate the course of battle, and thus to induce errors in deployment or other decision making. As such, traditional visually obscuring techniques such as introducing fog or smoke into the battlespace are now considered classic non-lethal sight weapons.

The primary effort today is to develop temporarily blinding lasers (Hecht 2012), the problem being that not all people react to light stimulus the same way (what temporarily blinds one may permanently blind another). These are some of the simplest non-lethal weapons available today. After classroom laser pointers became ubiquitous, numerous reports from airline pilots being lasered by what appear to be individuals out for a prank illustrate the safety problems that can result when unthinking individuals play around with high-tech gizmos. Blinding the operator of an opposing weapons system may not make that person stop shooting, however, though it will definitely hurt his or her aim (which is not good for the effort to limit collateral damage). If the opponent is operating a vehicle of any sort, he or she may not be able to avoid crashing into something or someone, a possibly undesirable intent of such an action. With more powerful lasers – and these are available commercially as well as in restricted defense laboratories – the victim could be permanently blinded. From a harsh military perspective, permanently incapacitating the enemy's troops may put a heavier burden on them than simply killing them, and could be justified as militarily expedient. The enemy must still care for its wounded, with no hope they will return to battle, draining resources that might otherwise be available for active combatants: "a dead soldier is just dead, but a blinded one needs the help of others, thus tying up several enemy soldiers at once – similar to the thinking behind the use of landmines to blow off legs and arms" (Drollette 2014).

Blinding is the most obvious use of directed energy in an anti-personnel role, but numerous other capabilities exist. An alternative visual weapon design is to produce strobe effects that could induce epileptic convulsions in those predisposed to them, but at least headaches and other temporary distractions ranging from disrupted decision making up to effective paralysis for the rest. The opponent is tired, increasingly irritable, and gradually demoralized as sleep deprivation takes its additional toll. When quicker results are needed, the same logic informs the use of flash-bang grenades by military and police Special Weapons and Tactics (SWAT) teams. These produce an intentionally blinding burst of light along with an ear-splitting noise to disorient and confuse belligerents in a hostage situation. Physical harm can come to an opponent who is too close to the grenade when it goes off, though vastly less so than a concussion- or shrapnel-producing grenade.

The big issue for sight weapons is that the amount of energy necessary to temporarily blind an individual is dependent upon the target's sensitivity, distance, whether the target is stationary or moving (and how fast and in what direction), and most important, the power generated in the laser. For these reasons, should the operator over-estimate the resilience of the target, making sure the minimum damage is achieved though the effects may not be temporary, or under-estimate, allowing for possible null effects to protect the opponent? The use of these weapons remains extremely controversial, evidenced by the 1995 UN Protocol on Blinding Laser Weapons (1995) to limit the use of these weapons as much as possible in future conflicts.

Sound weapons

Sound weapons can cause a variety of unpalatable effects. One such weapon, styled high-intensity directed acoustics (HIDA), purports the capacity to produce a sonic "bullet" as effective as physical ones in that they create a pain so intense as to be disabling. It also mimics nerve gas in that it produces disorientation, dizziness, nausea, vomiting, itching, fainting, and even migraine symptoms. "If you stand in the beam for more than 10 or 12 seconds, you get sick. People turn as green as grass, and you can pulse it in such a way that their ears don't really recover – so they keep getting this uncomfortable effect and they can't brace themselves to get away from it" (Weinberger 2008). When added to high-intensity strobe lights capable of producing epileptic-type symptoms, the combined effect can be astonishing.

For the most part, sound weapons are designed for crowd control and dispersal and are focused on very low-frequency transmissions. The sound is so low, in fact, that it cannot be heard by humans. But it is capable of causing all manner of distressing physical behavior. It is also possible to simply pump very loud noises to induce irritability and reduce efficiency. The sound of a baby crying, played backward and at intense levels, is especially effective, but even common (and to some quite pleasant) sounds have military application. When the US invaded Panama in 1990, it directed continuous, extremely loud heavy metal music at the Vatican embassy, where Manuel Noriega had been granted refuge, as part of a psychological operations package (McConnell 1991). The use of music has also been used as a form of torture, and in some places classical music is used to discourage teenagers from hanging out.

Smell weapons

One category of hyper-sense bombardments to offer alternative reversible or non-lethal effects is olfactory weapons – the classic stink bomb (also known as malodorants). If the smell is bad enough, it can flush enemies from their concealed positions. Combined with psychological operations, smells can be associated with culturally offensive symbols and enemy combatants can be painted with foul odors that make them the object of derision or even tagged as outcasts in their own societies. Some such weapons make an enemy combatant smell so foul – and if injected into the physiognomy of the person, the smell is extremely difficult to wash out or remove – that he or she is ostracized from the group.

Israel has already deployed a malodorant crowd control system called Skunk, for the smell it resembles. It has also been described as "An overpowering mix of rotting meat, old socks that haven't been washed for weeks – topped off with the pungent waft of an open sewer" (Davies 2008). At any rate, Skunk is sprayed from a water cannon mounted on a van to disperse crowds. The smell is extremely resilient; it takes at least three days to scrub from the skin, and up to five years to completely eliminate from clothing.

Of interest, the most offensive odors seem to be human-specific, and are somewhat culturally specific, too. Fecal odors, for example, are less bothersome in regions that do not have routine indoor plumbing and where the smell of human feces is more prevalent. The most typical malodorants are therefore feces, sewage, burning hair and flesh, and vomit. Another way in which offensive smells can be used to manipulate is in standard aversion therapies and interrogations.

Burning weapons

In 2010 the US military fielded its Active Denial System (ADS) for crowd control and selective operational use. ADS employs a high-output microwave beam that can be broadly distributed for group or area denial, but can be narrowed to a beam capable of targeting a single individual from up to a kilometer away. The system uses a two-second microwave burst that penetrates human skin 1/64 of an inch, giving the sensation that one is on fire. Although the skin is actually heated up to 130 degrees Fahrenheit, there appear to be no lasting effects, and the burning sensation stops when the microwave is turned off. Numerous dignitaries and reporters have been offered the opportunity to experience the beam, and many have done so. None were able to tolerate the beam or prevent themselves from jumping and raising their arms, usually dropping whatever they were holding.

Since the intense burst of pain is short, it is meant to disrupt or stop an action only for a moment. Crowds disperse because the pain can be repeated as often as necessary, and even continually if the targets use various protection measures. This is, of course, not useful in every case. If some of the targets are protected, or are particularly resilient, the operator of the ADS can increase both the strength and duration of the microwave beam. The normal strength doubled to four seconds of exposure can cause blistering on the skin. Longer and more powerful beams could be fatal.

Fortunately, ADS has other very desirable military capabilities. Very strong microwaves can be used to detonate improvised explosives and even minefields, and so ADS at the front of a column of trucks providing military supplies or humanitarian relief could reduce the likelihood of a roadside bomb interfering with the operation. Of course, the microwave beam powerful enough to detonate an explosive is too strong for crowd control. High-powered microwaves are also being studied as an anti-missile weapon, as they can disrupt onboard targeting sensors and computers, causing the missiles to go off course.

Electromagnetic weapons

A class of directed-energy weapon that is highly promising is high-powered electromagnetic (EMP) and microwave pulse weapons that could fry all modern electronics that use semiconductors. Specifically designed to take down electrical grids, power generation, and information systems, the weapon uses a very short, intense burst of energy to overwhelm electrical capacity and effectively short out equipment. Such focused pulses could be used to incapacitate large areas of electronic activity, perhaps up to a city-block or so in radius. They would also be effective in exploding munitions from a distance, perhaps very useful in clearing minefields or improvised explosive devices (IEDs).

Another directed-energy weapon under development is the pulsed electromagnetic projector (PEP). This system fires an electric laser that is allowed to bloom (spread out), creating a plasma cloud when it comes in contact with an individual or physical impediment that is strong enough to knock persons off their feet and render them unconscious, leaving pain, nausea, and dizziness as aftereffects (Beason 2005). While more physically arresting and longer-lasting than the ADS, it could also have less potential for misuse or unintentional effects, as the pulse strength could be fixed at a less than lethal level. Still, it is highly contentious. In direct reference to ADS, Andrew Rice, a consultant in pain medicine at Chelsea and Westminster Hospitals in London, argued: "Even if the use of temporary severe pain can be justified as a restraining measure, which I do not believe it can, the long-term physical and psychological effects are unknown" (Hambling 2005).

Adventurism or deterrence?

Non-lethal technologies are not a panacea. They are relatively unrestricted today but will no doubt become more constrained by international convention and treaty. Some of the problems are obvious, and will be rectified quickly. Tear gas, for example, routinely used for crowd control *domestically*, is banned by the Geneva Convention from use in *war*. The same is currently true for using lasers to blind an opponent – quite illegal in war, even if temporary. Moreover, the rapid proliferation of these technologies will give rise to effective counter-measures. The microwave-emitter that causes a burning sensation penetrates normal clothing but not a light metal screen or mesh (a modified Faraday cage) that could be used as a shield or worn over clothing. Ratcheting up the output of the microwave would seriously harm the unprotected, and if alternative weaponry is not available, the non-lethal force could be overwhelmed.

Non-lethal weapons do provide an additional level of escalation, a tactic between doing very little and declaring full-on war that is precise, selective, and, in the right conditions, extremely effective. They should therefore be part of any modern military's arsenal. They should not be over-used or over-relied upon, however. Especially for *early* intervention options in areas of humanitarian concern or as an attempt to reduce a larger outpouring of violence later with a small injection of violence now, non-lethal force could permit action without significantly increasing the risk of escalation. In combat or warfare, it may also be useful when the opponent uses non-combatant civilians as a shield. Temporarily incapacitating everyone allows easier isolation and identification of troublemakers. It is also probably desirable to use some technologies, especially EMP energy, to disable electronic equipment and disrupt the opponent preparatory to an attack. Currently, kinetic air strikes destroy radar and communications facilities prior to an assault.

Non-lethal technologies sound nice. They promise to limit death and even pain (in the long term). And while they may be extremely useful in crowd control situations, effectively revolts or rebellions that reflect a current unrest but have no long-term political goals, they probably will only increase anger and heighten opposition by groups looking to claim governing legitimacy.

Indeed, the military forces of liberal democracies ought not to be the tool of choice for domestic unrest. They are not police, and they should not be. Likewise police should not use military equipment and tactics. Police and military forces have different objectives; the former limits violence while the latter projects and maximizes it. No organization can do both at the same time and do them *well*.

The larger problem for a state whose military employs non-lethal weaponry routinely is that it is unlikely to deter a committed opponent. Such tactics are annoying, even disruptive, but they do not quell the *reason* for the opposition and so only harden opponents. At the same time, because killing and collateral destruction are eliminated, decision makers may feel it is easier to *use* forces so equipped, and it could lead to a national adventurism in international matters. Where decision makers today may be reluctant to use military force in situations where the opponent is embedded with noncombatants or is operating entirely within the borders of another sovereign state, they may be less so in the future. In the end, while non-lethals may reduce the pain and destruction of war, they may have the undesired effect of making war more common.

Perhaps this is the price we pay for trying to make war humane. Maybe it isn't supposed to be nice, easy, and safe. All of the technologies described so far are already deployed or in late-stage development. But what is in serious consideration for the next iteration of science in support of the military power of the state? What is the future of war?

FIVE

What Will Tomorrow's War Look Like?

"The most important thing you need to know about the Pentagon is that it is not in charge of today's wars but rather tomorrow's wars."

Thomas Barnett

An American infantry sergeant prepares to take her squad on routine patrol. She slides a titanium and Kevlar exoskeleton over her uniform and adjusts the menu for normal-speed, heavy-lift operation. She'll be in a variety of rough terrain today – she walked this route several times on a virtual terrain simulator and knows what to expect – and wants all the strength her servos can muster. She pops in her Internet-accessing contact lenses and connects to the military network overlaying the planned battlespace. The access port embedded in her brain and connected to a receiver array on her uniform has more bandwidth, but she always puts in the contacts just in case. Information redundancy is highly valued in a firefight. She selects from among her med-adjust pack the supplements she will need to enhance her already finely honed, DNA-enhanced, and surgically augmented combat skills, and then dons her most prized and valuable weapon – her battle helmet. It is custom-fitted, laser-molded to only her head and neck. The interior is a concussion-blocking polymer that connects her unique brainwave patterns to an embedded communication and situation-awareness software suite. Tuned only to her unique brainwave pattern, any information she needs, as well as complete control of her servos, is available with a thought – and the data flow is bidirectional. When she is wearing it, the information she collects through her eyes and ears as well as from billions of electromagnetic and biometric nano-sensors coating her equipment and coursing through her body is transmitted, processed, and stored, providing continuous updates and intelligence to operators and control monitors at a safe base up to half a world away. She has the finest martial proficiency that science can muster; she is a cybernetic killing machine.

And she is not alone. Every movement is tracked by satellites in orbit and a dizzying array of remotely operated and fixed sensors. Autonomous battle drones stalk the landscape in search of targets pre-selected in friend-or-foe identification suites, swarming like bees or fire ants with ruthless efficiency when a threat is detected. Micro-drones are littered across roadways and paths ready to snag passing vehicles with petroleum-eating enzymes or cling to human-organic movements, then infecting their hosts with energy-sapping designer maladies. A variety of directed-energy weapons from high-power free-electron lasers to plasma and microwave bursts are available from fixed and mobile land, sea, and air platforms. But the acme of combat is the remotely controlled robotic vehicle (RCRV). These sturdy and heavily armed machines can automatically react to threats with speed-of-light response times, just like the swarming drones, but whereas the latter are limited by strict rules of engagement, the RCRV has a human operator that can override pre-programmed safety rules and creatively engage any target the remote controller can justify. The enemy adapts swiftly, and this type of unit can adjust to rapidly evolving perils and unanticipated environments – but unlike the American sergeant above, the human operators are far enough from the battlespace not to be personally at risk. Only the machine is damaged or destroyed by a poor decision or more powerful foe. The operators gain critical combat experience and live to fight another day with a fresh RCRV. This makes them extremely bold, and very, very deadly.

This is the vision of war in the future. It is not so strange as to be unrecognizable, and many of the technologies described are already operational or in prototype testing. It is also not so removed from war in the present...

Future fears – the most perilous technologies

The beginning of the next war will look remarkably like the end of the last one. Already, Predator-type RPVs (remotely piloted vehicles) fly silently overhead, clandestinely watching designated targets (human beings) as they move across the landscape, transmitting live feeds to controllers up to half a world away. The target won't be serviced (military term for killing or destroying) right away, even though a companion Reaper RPV with sophisticated missiles is available. Intelligence is being collected and collated and no engagement will be called for unless friendly assets (human or property) are directly threatened. Resembling more of a police stakeout or even sting operation, the goal is to find and destroy enemy forces higher up the decision chain. Who do the targets talk to (follow them also); where do they go; what are their habits? Satellites in orbit scoop up every electronic transmission. Cellphone and email communications are continuously monitored and are merged into a sophisticated database for deep exploitation. In its most recent wars, this is the default operational pattern for US forces.

Soon enough, the target is depleted, from an information perspective, and the order is given to take the individual down. There are a variety of ways to do this, and much depends on the impression the attacker wants to leave with associates or fellow combatants. Would a stealthy night-insertion of a SEAL team send the right message, or would a Tomahawk Cruise Missile in the middle of the day be best? Should the strike be invisible or public? Which would cause more fear? Which would cause more complacency? These are all contextually influenced, of course.

What we can say for certain is the next war, so long as it fits our definition of war and is not simply a tag to give the impression of serious engagement (e.g. war on poverty), will be violent, deadly, and destructive. In part because unfettered scientific inquiries end wherever an investigation leads, new technologies are systematically adapted for war. Despite the current emphasis on temporary, reversible, and non-lethal effects, the weapons of tomorrow will change the character of war immensely, and no one can predict the direction of technology or war as science continues to reveal the inner workings of nature. But it will still be war. The lessons learned from the last engagement will be foremost in the minds of planners, and the default will be to continue doing what one has been doing well. In peacetime, military decision makers will resist the more radical technology science offers in favor of gradual inclusion of new technologies that enhance existing capabilities.

As war drags on, however, the pace of scientific technological development quickens, receives more resources, and gains insight into emerging tactics and capabilities of the opponent. As shown in the historical examples from chapter 2, far-reaching change in the character of war tends to take place in the prosecution of extended warfare, not in extended periods of peace, and predicting sweeping change is always fraught with error. With this caveat in mind, the following are examples of so-called game-changing technologies that are receiving enormous scientific attention, and will impact not just the character of war, but the social and political rules that define our civilization.

War on demand: 3-D printing and the hyper-empowered individual

A few years ago, three-dimensional (3-D) printing was an afterthought of futurists. Today, it is one of the most promising industrial technologies as well as one of the scariest from public protection or national security perspectives. 3-D printing is an additive manufacturing process in which a digital computer design (CAD) model is built in real space one thin layer at a time. There are different means to produce these layers, some only an atom thick. Most commercially available printers do so by laying down slices of liquid, melted, or softened material one on top of the other, individually curing layers (or the entire object once built), and swapping out various base materials as needed (Lipson and Kurman 2013).

Most printers are still fairly small. To build a large object, individual components are printed and then fused together. In this way, enterprising individuals have managed to print fully functional automobiles, satellites (that have been launched into space), and water-going craft. Some are even experimenting with printing edible meals – researchers in the Netherlands printed a steak that was nutritious and (reportedly) tasty (Moskvitch 2013). The potential of having a type of base food component – I'll call it soylent for all the skeptics – that could print up a variety of tasty recipes portends a Star Trek-inspired replicator capable of making "tea, Earl Grey, hot" in a few seconds. Depending on the source – say algae – of the basic soylent material, widespread hunger could be eliminated. For military forces, it is reminiscent of the introduction of preserved (canned) food that soldiers would carry with them on campaign (see chapter 3), extending their combat range and allowing large armies to move through civilian areas without impoverishing them by foraging. The 3-D printer may eventually replace the camp cook.

For science, industry, and militaries, the printer is a logistical boon. While some supplies will have to be kept in massive warehouses and depots – for the military, anything that is expended in quantity in a fight – manufacture may be heading the same direction as media: print on demand. It will no longer be necessary for the military to maintain massive reserves of components. When a new filter or wheel hub is needed, simply load up the program and print it out. This is already being done for high-priced components of specialized vehicles including bicycles, motorcycles, and race cars. As the price comes down for printers and essential "inks," businesses and militaries will have very few warehouses. Moreover, *nothing will ever go out of production*. If you like your old Kelvinator kitchen appliance, somewhere on the Internet will be that compressor part you need. Push start on your microwave-sized printer. Then get on YouTube and watch the instructional video as you install the piece. Military logistics will transition from specialists (e.g., helicopter mechanics, equipment repair, bomb disposal) to generalists capable of near-instant expertise, so long as they are connected to the Internet and a printer.

There will be many other changes, most of which are yet to be identified. Already, prototypes are fast-tracked and tested, bringing new products to consumers and allowing previously financially prohibitive incremental improvements to old ones to become viable. SpaceX Corporation is already using printed parts on its cargo carrier rocket, and its planned human-rated space transport vehicle for service to the International Space Station, the Dragon V2, is partially printed, including much of its rocket engine (Bergin 2014). NASA will soon launch a 3-D printer to the space station so astronauts can make replacement parts onboard instead of having to wait for a resupply ship or maintain a large inventory. Dentists are creating crowns and full dentures for their patients while they wait. Optometrists are printing glasses – and soon contact lenses – in their offices. Medical doctors and technicians are printing a wide range of prosthetics, but also implants, to include genetically compatible bones and ligaments. They are also experimenting with printing body parts to include viable organs, bio-coded to resist rejection. The Chinese have already printed ears, kidneys, and livers. British scientists are working on printing chemical compounds, looking to the day when family physicians will simply print out the medications they would otherwise prescribe.

But what could be a serious and positive social and military change agent also has its dark side. What could be printed that could be dangerous? Materials that already can be used to print objects include high-density polyethylene plastics (where corrosion and chemical resistance is needed) and thermoplastics (very hard substances that soften when heat is applied and return to their original shape and hardness when cooled); paper and plaster; clay, porcelain, and metal-ized clay (for ceramic applications); silicone rubber; almost *any* metal or amalgam to include stainless steel and titanium. It may soon be possible to print out high explosives as easily as a beef stroganoff. An all-plastic handgun could be printed out today that shoots plastic bullets, all invisible to metal detectors. Improvised explosive devices (IEDs) could become the weapon of choice for any disgruntled worker as well as committed terrorists. If chemicals can be printed, individuals may create dangerous poisons or gasses, even biological diseases that could have massive deadly effect. It may even be possible to print an atomic bomb. In your kitchen.

What happens when an angry, isolated person decides to strike out at the world – Thomas Friedman's "super-empowered individual" who happens to own a printer that is limited in its creative capacity only by the intelligence and hatred of the user (Friedman 1999)?

Some controls on raw materials will have to be enacted, as will limits on the kinds of CAD blueprints allowable on the Internet – but these will likely only be hindrances. Let us hope that the creative scientific genius of altruistic individuals will invent protective or preventive capabilities faster than antagonistic ones can devise deadly new weapons. Regardless, science is leading the military into profound logistical changes that present both opportunity and peril. The need to explore the universe, to know the natural world precisely, without regard to moral or ideological imperative, now moves from the macro world of printers and the Internet to the micro world of germs and nanoparticles. The military will adapt.

Micro war – it's the little things that'll get you

Although chemical and biological agents are prohibited under international treaty, their use can only be expected to increase as more non-state actors – undeterred by threats to territory or sovereignty – become more troublesome, creating an entirely new arena requiring military protection. The reason is simple. The potential damage that can be caused relative to the effort expended to cause it is enormous. In an asymmetric conflict, the less powerful will opt for the most terrifying threat at the lowest monetary cost. Enter the realm of biological and chemical engineering.

Much of the fear of micro and nano war is focused on designer diseases that could be surreptitiously introduced into the atmosphere, food supply, or community water systems and cause massive casualties and follow-on stress to the national infrastructure. Many of these are expected to be human-designed or -created diseases and viruses, though they could be known pathogens recreated in the laboratory. Indeed, rumors persist that AIDS and Ebola were created in CIA or KGB laboratories specifically targeting minorities with identifiable genetic traits. There is no hard evidence to prove this, and it seems much more likely that AIDS, which didn't exist 40 years ago, came about through some interaction of over-use of antibiotics, environmental pollution, overpopulation, and mutations allowing it to jump from animals to humans in an unpredictable way. Still, it is quite possible to create designer pathogens in the laboratory. The easiest way is to take an existing virus, such as polio, typhus, or flu, and make a small adaptation to it. Nature already does this through mutation, which is why so many common diseases are increasingly resistant to drug treatments, and a first-year graduate student could do it armed only with a map of the human genome (freely available on the Internet) and a basic set of chemistry tools (Mauroni 2007; Paxman and Harris 2011). And the worst part is one doesn't have to be intent on destroying humanity to make a deadly new virus – some are simply byproducts of well-intentioned schemes to do good that had horrifying side effects. The use of thalidomide, for instance, in the early 1960s as a calmative that provided effective relief from the symptoms of morning sickness in pregnant women was subsequently found to produce terrible birth defects.

Nor does one have to invent an entirely new disease to create a massive human tragedy. Smallpox, for example, had been effectively wiped out through a global immunization program that ended several decades ago. With the disease dormant, immunizations were discontinued, and the great majority of the earth's population may have no smallpox resistance. Smallpox is extremely virulent, and should it re-emerge would likely kill up to 40 percent of those infected. Food poisoning, caused by *Salmonella* or *Botulism* bacteria, is deadly and easily acquired. It has been spread intentionally in several places, including the first known bioterrorism attack in the US when a religious group infected salads at a dozen restaurants in Oregon and sickened more than 750 people. Fortunately, there were no fatalities.

A potentially positive use of biotechnology is modification of the human body through DNA manipulation (Shanks 2005). Injecting extra copies of genes that stimulate the production of endorphins, for example, could reduce the effects of pain and stimulate endurance. The same could be done for testosterone or estrogen production. Genetic splicing goes a step further, actually inserting genetic traits from one species into another. Using this technique it may be possible to enhance a human's eyesight to the level of, say, a bird of prey. Skin could be modified to add protective scales or the toughened hide of an alligator. Some individuals could learn to navigate in pitch darkness through introduction of genetic codes responsible for a bat's sonar. Such capabilities are still a few years from routine use, but the theoretical and technological requirements for doing so have already been demonstrated.

Another way to biotechnologically modify human traits is by inserting specific drugs into the body attached to a virus (usually the common cold; see Savulescu et al. 2009). One such drug, ibutamoren, increases the production and release of growth hormones responsible for muscle mass and bone density and is currently under development to help growth hormone-deficient children and elderly persons with osteoporosis. It also tends to reduce body fat through manipulation of metabolism, making it a potentially valuable anti-obesity drug. And if adding hormonal power in some cases is tantamount to improving human performance, so is decreasing it in other cases. By disabling a naturally occurring protein called myostatin, scientists have observed the doubling of muscle mass in mammals. The hope is that a drug capable of blocking myostatin could help in the treatment of muscular dystrophy and other muscle-wasting diseases, but what warrior (or warrior's boss) could resist gaining twice his or her strength and endurance?

Nanobiotechnology, the science of developing nano-sized (atomic scale) machines that can be injected into an organic body to alter its basic characteristics, is a potentially powerful way for scientists to develop artificial antibodies for designer diseases, correct genetic malformations, and enhance performance, and many, many other applications (Ach and Lüttenberg 2008). Like the problem of containing an infection or poison that is readily transmitted, nanobots have been studied for their resistance to counter-agents. The reason is simple enough. Any introduction of a trait into an organic body that is beneficial to that body is capable of being replicated in the gene structure of that body and passed on to successive generations. How that occurs, and whether various mutations that attend generational characteristics transmission are also beneficial, is still relatively random. This is especially the case when instead of injecting millions (even billions) of nanobots into the organic system, it may be more broadly effective to inject a few nanobots that are capable of reproducing themselves. In 1986, Eric Drexler was attempting to extrapolate scientific developments in the near future and brought up the possibility that self-replicating nano-machines could overrun the planet. The process is simple. Nano-machines would take in organic matter and transform it into fuel and building material to create new nanobots, potentially *ad infinitum*. In this vision of microtechnology run amok, the entire world is eventually transformed into "gray goo," a sickly syrup of biological waste material that ends up covering the planet. The good news is that non-self-replicating nanobots are extremely cheap to produce and technologically much easier, so it would take a mad scientist *intent* on destroying the world to actually go to the bother of building gray goo capable nanobots... and how likely is that?

Currently, nanobots are some of the promising vectors for treating cancer, anemia, asthma, immune system disorders, and the like. But they have their downsides, especially in an increasingly miniaturized world in which it is widely assumed that within the next generation *most* people will have some form or another of artificially created nano-machines coursing through their veins and tissues. The benefits are simply too good to forgo: less pain, more endurance, better hearing, etc. A larger question comes up when modifying specifically human warfighters or wartime decision makers with enhanced physical and mental capabilities. A housecat, for example, that has increased eyesight and faster reflexes will likely kill more small birds and rodents than it had previously. Humans that are genetically or tech-nologically modified to be superior warfighters will tend to find themselves in command positions that give them increased authority over non-enhanced persons. They may even begin to perceive themselves as superior to the non-enhanced, and may seek to subordinate or weed them out over time. We simply cannot know. Any effort to use micro- or nanotechnology to reduce aggressiveness in humans as a means to ending war – discussed in the previous chapter – *could* put them in a seriously weak position relative to the aggressively enhanced (or even the remaining unenhanced) humans.

Robot war – no pain, all gain

In 2000, Japan's Honda Corporation unveiled ASIMO (Advanced Step in Robot Mobility), the most advanced autonomous humanoid robot at the time. Child-sized at about 130 cm in height, ASIMO has been upgraded significantly since its debut, and in 2014 played a bit of soccer with President Barack Obama (Kaku 2011: 68). ASIMO has two visual sensors where a human's eyes would be, allowing it to determine sizes and distance of objects, recognize objects and track their movement, and, important to its design, appear more human-like. It is also capable of facial recognition and taking cues from human posture and gestures. Similarly, its auditory sensors are located where a human would expect ears. ASIMO can distinguish sounds to identify individuals and, with its eyes, appears to follow conversations and turn toward noises. It has also been programmed with its own body language, appropriate to an adolescent human, to include jumping in place to show excitement or impatience, offering handshakes when appropriate, and encouraging humans to play games.

While ASIMO shows the level of development in robots designed eventually for household use, it also has helped pioneer numerous production and military robots. The most widespread use of semi-autonomous military robots today is for remotely piloted vehicles (RPVs), and the most famous of these is the US Predator series aerial vehicles (Springer 2013). Predator is a long-endurance, medium-altitude unmanned aircraft system for surveillance and reconnaissance missions, first developed by the CIA. Now used by all America's national services, imagery comes from synthetic aperture radar, video cameras, and a forward-looking infrared (FLIR) system that can be distributed in real time both to the front-line soldier and to the operational commander, or worldwide in real time via satellite. The MQ-1 variant is armed with a pair of AGM-114 Hellfire missiles, allowing Predator to accomplish additional military missions, including interdiction and destruction of high-value or unexpected targets. An MQ-9 Variant, called Reaper, can carry up to four Hellfire II anti-tank missiles, two laser-guided bombs (GBU-12 or EGBU-12), and a 500lb GBU-38 JDAM (joint direct attack munition). In addition, Reaper carries a synthetic aperture radar and has a ground moving target indicator.

Incorrectly called drones by most, the RPVs in use in militaries around the world are essentially advanced remote-control model aircraft with semi-autonomous flying capabilities (much like an autopilot) and system management computers. Future variants will have extended range, speed, and carrying capacity and will be expected to conduct near-autonomous patrols. Armed with weapons, the technology will come from autonomous fire suppression systems already in development that include unmanned firing turrets for installations – requiring intruder detection capabilities matched to targeting and weaponry – and automatic targeting and gunfire for helicopters and low-flying manned aircraft. The ability to detect and react to a threat is already approaching the limits of human cognition and response, and so such automated firing systems are supplements that can be overridden by the operator, but would generally default to autonomous operation in highly dangerous locations where friendly assets are not expected.

Air power has benefited immensely from RPVs, but military interest in more autonomous robotics has skyrocketed across all domains; land, sea, air, and space. The advantages are clear. Robots don't need sleep, and don't tire. When humans get shot at, they tend to get angry and shoot back. Robots aren't so emotional (yet), allowing for strategic restraint in dangerous but fragile situations. And as they become more ubiquitous, they become much, much less expensive – especially in comparison to the life and limbs of a human being. They are disposable.

With that in mind, some of the robots already deployed today include Big Dog, a four-legged large-dog-sized robot that can scamper across heavy terrain inaccessible by wheeled vehicles (it can cross rubble and climb most stairs) and is capable of transporting up to 200 kg of cargo. Essentially a mechanical mule, Big Dog accompanies human patrols packing extra munitions and supplies. Upgrades include Alpha Dog, larger and capable of walking on ice and other slippery surfaces, and in 2013 Big Dog was fitted with an arm that can pick up objects weighing 20 kg and throw them with decent accuracy.

Military robots are ideal for dangerous missions, and early uses are bomb identification and disposal and urban reconnaissance. Simple remote-controlled robots such as the US MARCbot (Multi-function Agile Remote Control Robot) and India's more sophisticated Daksh are small wheeled platforms with cameras that can inspect under automobiles or around corners. Daksh has a robot arm that can carry or retrieve items, an x-ray sensor to detect hidden bombs, a shotgun to open locks and other barriers, and a powerful water jet that can defuse or destroy bombs too dangerous to move. Packbot is the US military's multipurpose small robot. It can be outfitted for reconnaissance, bomb disposal, medical support, sniper identification, and hazardous material detection. Future generations will have the ability to mount rifles, machine guns, anti-tank weapons, and grenade launchers. All of which brings pause to those who envision fully autonomous killing machines roaming the battlespace looking for mechanical and organic targets.

In November 2012, the US Department of Defense issued Directive 3000.09, "Autonomy in Weapons Systems," to address many of the concerns related to computer-generated decisions to kill or maim or to damage property. Its stated purpose is to establish policy and assign responsibility for the actions of autonomous and semi-autonomous weapons systems, as well as develop plans to minimize the probability and consequences of failure in these systems that lead to unintended effects. Although it also establishes relatively easy bureaucratic processes to change the rules, to date the directive prohibits lethal, fully autonomous robots and limits semi-autonomous robots to engaging only previously selected targets or those selected by an authorized human operator (Enemark 2013). The Directive followed a July 2012 Defense Science Board Report that took the US military to task, stating that "autonomy technology is being underutilized [due to] obstacles within the Department [of Defense] that are inhibiting acceptance of autonomy and its ability to fully recognize the benefits of unmanned systems." These obstacles include "poor design, lack of effective coordination...and operational challenges caused by the urgent deployment of unmanned systems to theater without adequate resources or time to refine concepts of operation and training." When you don't know where you're going, don't be surprised where you end up.

There is little doubt that the future of high-technology warfare is tied to robotics. Moreover, it is enabled by massive scientific breakthroughs in communications technology. Remote-controlled military platforms are supported by global secure communications that are heavily reliant on spacecraft for observation, intelligence, and command and control of the remote platforms. Operators are far removed from the battlespace and so have different reactions to war than those in the line of fire. On the one hand, they may be more likely to engage an enemy as they have no fear of personal harm. On the other, they may be more restrained, as the normal human reaction to being shot at is to get angry and shoot back. A certain detachment can be expected, unique in the annals of war, from combatants by day who drive home every evening to spend dinner with the family.

From the micro to the macro, science is enabling a new kind of warfare that military planners are trying desperately to understand and then incorporate into their strategy and tactics, all of it coming back to the revolution in military affairs augured by the massive rise in storable and searchable information. The vector of influence is two-way, of course, but the initial work was partially pioneered and funded by military necessity. Following the launch of Sputnik in 1957, President Eisenhower authorized the creation of the Defense Department's Advanced Research Projects Agency (DARPA), tasked with developing promising new technologies that could result in breakthrough military capabilities with further commercial or civilian applications. In 1968, acting on a USAF Project RAND report detailing a need to connect the nation's primary nuclear weapons control computers in the Pentagon, North American Aerospace Defense Command (NORAD: Cheyenne Mountain, Colorado), and Strategic Air Command (SAC: Omaha, Nebraska), DARPA funded a promising packet-switching network concept proposed by computer science engineers at the

University of California-Los Angeles. One year later, a proof-of-concept network dubbed ARPANet established a link between UCLA and Stanford University. By 1970, nine universities and commercial laboratories were connected. Eventually, the system grew into the World Wide Web – today's Internet.

Information war: Big Data and you

It may not seem like a particularly frightening or scary component of future warfare, but the Information Age has spawned a form of information collection, processing, and analysis that portends serious threats to the expectations of personal liberty and privacy that many in the developed world take for granted. The massive amount of information that has been stored in the last two decades has created an enormous database that is subject to sophisticated, highly complex analytics. The information coming out of the new focus on Big Data is surprising – both extraordinarily helpful and worrisomely life-changing. And it all originates in wartime technological development.

Modern warfare is information warfare, and intelligence is what the military calls information. The side that has better intelligence has a distinct advantage over its opponent, especially when that opponent is unaware of, or cannot determine, what information the other side has. In the past, when intelligence was scarce and hard to come by, military commanders preferred highly classified information that was *unavailable* to others. For this reason military communications were tightly controlled, heavily encrypted, and limited to highly trusted agents. When allied code breakers decrypted the German ENIGMA cipher in 1940, the fact that they could read the enemy's most sensitive transmissions was an enormous boon that at a minimum shortened World War II and may have won it. ULTRA, the program name for the Allied code-breaking effort, was the most successful story of intelligence and counter-intelligence warfare in a world of scarce information (Winterbotham 1975; Budiansky 2000).

By the 1990s, however, military intelligence capabilities, especially electronic and communications collection, had become so good that more information was being gathered and processed than could possibly be individually analyzed. In this world of abundant information, intelligence practices that once worked brilliantly were becoming less effective. News organizations and Internet bloggers were coming up with militarily significant information faster and with greater reliability than defense analysts could write up their reports. Old ways die hard, especially those that were once extremely effective. When information is scarce, hiding it away makes sense. When it is abundant, why bother?

Enter the age of Big Data. The reason information is so abundant has to do with computers and digitization. Information that was once kept on paper and in secure locations is now in computer databases and large-capacity electronic storage, and most of that on-line. It is eminently quantifiable and searchable, but the amount of data is so staggering that computers have barely been able to keep up, and the pace of data collection is soaring (see Mayer-Schönberger and Cukier 2013; Davenport 2014). Mobile phones with cameras and video recorders are everywhere, most capable of linking directly to the Internet. Add to this traffic and weather cams, software logs, radio and television broadcasts, credit card readers, barcode and RFID scanners, and the amount of data added to the total every day is mind-boggling. The biggest contributor to Big Data is social media, led by Google, Facebook, Twitter, Amazon, Vine, Instagram, and LinkedIn, but also including picture and video dump sites such as Flickr, Reddit, and Imgur. Data is being generated every day, all around us, from multiple sources at a rate that is barely comprehensible to the human brain.

As the number of sensors increases, especially in urban or high population areas, we may be looking at the end of privacy as we know it. By choice (Google Glass, personal enhancements) or by habit (posting to social media, driving on public roads), it will be harder to find a moment when one can dally quietly unobserved. When it happens to billions of people, all the time, most will become comfortable with the notion and will start acting naturally. This is what happens, especially to younger people who have grown up with social media. They post the most personal information with nary a thought – pictures of their last meal with location enabled, relationships, vacation plans, sexual orientation, and more. They probably do so with the same sense of safety that a fish in a school experiences. When the big fish comes for dinner, it is more likely to snack on the first it finds and the rest happily continue on. Unless it is a really, really big fish. Then it might snap up the whole school; and this appears to be where we are today. We have an enormous mountain of data, filled with veins of misguided, inept, and inaccurate inputs, but when aggregated an extremely precise picture of *social* activity emerges. Individuals are swimming in this data sea with varying levels of concern for their digital privacy or safety, for as of yet the school of digital fish is very large and the predators are relatively small. As the state gains more capacity relative to the individual, and vice versa, the individual gains increased capacity relative to the state, the future of political violence will surely be affected.

So how does it all work? Currently, sophisticated software algorithms are being written for use on powerful super-computers to comb through Big Data and look for trends and anomalies that will allow forecasters to make accurate predictions about the future. For example, a company called PredPol (Predictive Policing) adapted a sophisticated algorithm for predicting earthquakes and input all of the digitized data from the Los Angeles and Santa Cruz police departments in an effort to predict outbreaks of criminal activity (Perry et al. 2013). The software has achieved surprisingly good results, and is reportedly able to predict where crimes are likely to occur down to an area of 500 square feet. Where the software has been tested, Los Angeles police are reporting a 33 percent reduction in burglaries and a 21 percent reduction in violent crimes. In another example, after noticing spikes of search activity just *prior* to major international events such as the 9/11 terrorist attacks, several Big Data programs have been looking for patterns that might help determine where and when a terrorist strike happens next.

But Big Data can also help in isolating and exploiting individual activities of high interest. Once an individual has been identified as a person of interest, computers can quickly comb through petabytes of data to isolate Internet activity, credit card purchases, emails and texts, banking data, police records, etc., all combining to provide a very accurate psycho-social profile. If the manipulators of Big Data want to destroy an individual, no one will be exempt; no one will be able to hide. Friends and colleagues through a lifetime can be identified and tracked; routine patterns of movement such as going to and from work can be identified and the target intercepted; likes and interests that could be used to coerce or entice an individual; fears that might assist an interrogation; and more – limited only to the imagination of the data scientists who are developing the Big Data exploitation processes and algorithms.

Ray Kurzweil has written of the coming singularity, which he identifies as the moment an artificial intelligence becomes self-aware (2005). The idea is that a single computer will someday have the parallel processing power inherent in the average human brain, and as the speed and rate of processing in computers increases, sentience will emerge. He isn't quite right. The singularity has already happened, and it occurred in the recent past when the amount of data collected and stored became so large that new, unanticipated properties of human and computer interaction spontaneously emerged in the form of a new kind of intelligence.

Science is about gathering and categorizing knowledge for the purpose of exercising control over nature. We want more stable weather, hedges against future sickness, new opportunities to grow and gain wealth. But there is an inherent distrust between the individual and the state that, in this case at least, is facilitated by the emerging sciences. All of the developments described in this chapter, ubiquitous monitoring and surveillance, robotic and human-enhanced enforcers ready and able to augment the state's political and military operations, distributed manufacturing, and highly digitized learning, coping, and integrating algorithms based on aggregate data over time will increase the ability of the state to monitor and control populations. On the other hand, increasing capabilities of individuals to resist the state, through enhanced information and processing capabilities brought about by access to militarily important information distributed on electronic media and 3-D printing, have the potential to change the social fabric of developed states in the decades to come. The state may perceive itself relatively secure while a seething digital underclass carefully hides its intentions until it is ready to strike. Or not. The world will be different in character – it is different already – but the essential logic of human interaction will remain, and the solution to innately humanized problems such as war will not be calculating science but emotional attachment and social preferences. In this world there are no solutions from pure science that will end war, only practical advice on how to make it more or less efficient and effective.

SIX

What *Will* End War?

"We set sail on this new sea because there is new knowledge to be gained, and new rights to be won, and they must be won and used for the progress of all people. For space science, like nuclear science and technology, has no conscience of its own. Whether it will become a force for good or ill depends on man, and only if the United States occupies a position of preeminence can we help decide whether this new ocean will be a sea of peace or a new terrifying theater of war."

John F. Kennedy

"I have cherished the ideal of a democratic and free society... it is an ideal for which I am prepared to die."

Nelson Mandela

We have seen that science, like war, may have its own grammar but not its own logic; its own method but not its own prudence. War seeks total victory without policy to limit it, a historically tragic error. Left to its own, science will seek value-free knowledge to whatever end it leads, freed from political limitation or societal mores.

In both cases, without an external guide, perversion is the result. In Vietnam, for example, it is said that America won every battle even though it lost the war. Unable to provide a rationale other than victory as the conflict stalemated, the US military developed a body-count mentality. So long as more Vietnamese were dying than Americans, surely it must be winning its brutal war of attrition.

Such pernicious logic has long plagued the military force that seeks victory as its purest goal. After being heartily congratulated for his impressive triumph over the mighty Roman legions at Asculum in 279 BCE, King Pyrrhus surveyed the loss of the bulk of his army and lamented, "Another victory like that and we are done for" (White 1995: 393–4). In this case victory became such a relentless and important military goal that the political reality of fighting an *ongoing* war, in which future battles are likely despite early or surprising tactical success, is a deficient strategy. Separated by almost 2,000 years, the logic is sustained. In both Asculum and Vietnam, the relentless pursuit of battle-field victory directly contributed to losing the war.

Similarly, in science the goal is always knowledge, and gaining *more* is the end that justifies the scientific means. Morality is eschewed as an artificial limiter, yet even when a moral or ethical goal is externally mandated, the result often goes against expectations. This is because science cannot evaluate morals or virtue except as external confounding variables. Since science accepts only *measurable* data as factual, if it cannot be observed and quantified then it is beyond the reach (or interest) of science. A thing cannot be made better, more effective or efficient, if its progress cannot be charted. The criteria for measurement becomes the *new* goal, and progress will surely be recorded, but at the expense of other factors not identified as controls. For example, chapter 4 included the assertion that the downside of tasking science to make war *safer* can have the chilling secondary effects of making war more *likely* and *longer lasting*, a condition that may be contributing to America's current entanglements in Asia. This is a paradox, a situation in which two equally valid statements cannot logically coexist together. There is no solution to paradox. One must simply move past it, leaving a residue that can haunt future outcomes.

If scientists are given the task of eliminating war, defined as violent conflict between groups seeking governing legitimacy, the most efficacious means might simply be to wipe out humanity – if there are no people there can be no war – or to create weapons so horrifying that the prospect of eliminating humanity makes any attempt at war self-correcting. The latter was thought to have been accomplished with the introduction of massive arsenals of nuclear weapons in the twentieth century's Cold War. To the extent that science is tasked with eliminating a certain *kind* of war or a certain *element* in war, in the process it defines and provides for the opposite outcome just as surely. When nuclear war is made unthinkable, for example, then conventional war becomes more palatable. Hence the problem of moral hazard exists wherever the directly measurable or observable-only logic of science competes with the hidden, emotional responses of human beings.

Such pessimism is no doubt disheartening to those who hope for a scientific solution to war. It should be, but it should not be taken as a call to end scientific interaction with war. Such would be absurd and, given the scientifically driven situation the world is in today, counter-productive. Science alone cannot end war, but people are unlikely to do so without enlisting scientists in the effort. So this book concludes with some intriguing possibilities for peace down the road that have emerged as a result of scientific explorations, even though peace may not have been the scientifically determined intent of the investigations. If science is not perceived as a solution, but as a critical intervening variable in a process *toward* a solution, an *indirect* influence in support of a moral, ethical, or politically determined social good (that is, peace), then at least one avenue of investigation and action appears rife for exploitation: continuation of hard science space exploration supported by social science-derived knowledge of human motivation.

Hard science and the potential of outer space

In chapter 1, Thucydides' enduring categorization of the causes of war – fear, honor, and interest – was outlined as a basic structure for assessing the utility of science in determining if and how an end to war might be supported. Using Machiavelli's hierarchical form – security, wealth, and prestige – the following highlights the potential of a renewed emphasis on space exploration and exploitation that could address and ameliorate all three sources.

Death from above: Security advantages of space power

Fear of loss is the most primitive motivator, and life is one's most precious property. At the highest levels of collective or social action, the state's first function is to protect its population from violence. As described in the opening chapter, the primary means of protection from external violence are (in order) national defense, deterrence, and offense. Defense is the ability to thwart or stop an attacker from achieving its objective. It is active in that fortifications, walls, fixed weapons, and other obstacles must be emplaced, populated, and maintained. It says to the would-be attacker, go ahead and try to take my property, you will not get what you want. Deterrence is a different concept. It says my defenses may not be sufficient to keep you from achieving your objectives, but the costs I will impose on you for taking them will be so great you will rue the day you even contemplated attacking me. Thus, deterrence is a *threat* of unacceptable retaliation should the attacker be so foolish as to test my resolve (Art 1980).

Deterrence and defense are complementary. A stout defense will be deterring to the defender, but the calculation of cost and benefit is entirely at the discretion of the attacker. The stronger the defense, the more likely the attacker would decide not to waste the time and effort. But defense is expensive. All those bulwarks must be built, maintained, and staffed. And if or when defenses break, the attacker wins. Deterrence, on the other hand, is comparatively cheap. The expense involved is minimal – diplomatic only – but the credibility of the threat is enhanced the more offensive or retaliatory power the defender controls and has made visible to the erstwhile attacker. Offense is at least as costly as defense, but has its advantages.

Offense is the backbone of a deterrence strategy. It is what makes the threat of retaliation believable. It is also the key to a successful defense. A defensive stance with no ability to counter-attack when the enemy has become weakened or overstretched is doomed *eventually* to fail. Enemies will simply keep probing, pressuring, and attacking until one plan or another eventually works. Time is on their side. Losses are an occasion to salve one's wounds and rearm for the next round. Accordingly, a primary component of offensive capability is compellence – the theoretical opposite of deterrence. Compellence is an action taken by one state that it refuses to stop until the other side complies with its demands. A blockade, for example, is an active measure that stops supplies from entering a city or port. The blockade is not lifted until the desired action is taken by the opponent. One can presume if the offensive action taken is significant, and the threat of even more punishment is held out, the target will comply: the greater the offensive capability, the higher the likelihood of compliance.

All of the primary forms of protection are interrelated. They are presented separately only to highlight the strengths and weaknesses of each. Total reliance on one is ludicrous in practice, and highlights the paradox of offense versus defense in isolation. To illustrate, the Chinese character for war is "spear-shield," and is derived from the tale of an itinerant entrepreneur who offered to sell to the king of one warring state an impenetrable shield that could be pierced by no weapon. Armed with the device, the king felt sure of victory. The entrepreneur then went to the opposing king with a miraculous object, a spear that could penetrate any shield. Delighted with his purchase, the second king went confidently off to battle.

Let's start with the idea of an offense so powerful that no state would dare go to war with the owner of such a capability because to attack is to be obliterated. This is the logic of the nuclear balance-of-terror from the Cold War. So long as no side can expect to attack without being destroyed in retaliation, why would any sane leader risk war? Unfortunately, in the Cold War this meant that nuclear powers had to accept *mutual* vulnerability as their price for peace. And of course, stability was based entirely on the sanity – or rationality – of a nuclear state's leaders, a sometimes dubious claim. If mutual deterrence of states armed with nuclear arsenals ever does fail, the result could be global suicide.

And there were two possibilities for failure. So long as the retaliatory or second strike was not fixed to a tripwire or automatic method (the so-called doomsday device), it was expected that a human being would be the final arbiter of the response. If the attacking state believed that person would not have the resolve to strike back, deciding to not launch a massive response and instead simply conceding, a first strike might just work. The second possibility was that one side or the other might develop an anti-missile capability in secret, or would get a sudden defensive capability from a scientific or technological breakthrough. In such a case, the side with the defensive strength could decide to attack the state still relying on deterrence for its security.

Are there weapons in space that would be so frightening, so offensively capable that no nation would dare attack another for fear of their immediate demise? Not yet. At least, not yet deployed, but the technologies to do so are already available. If linked to a sophisticated computer or artificial intelligence, perhaps one based *in* space where it was relatively immune from a first strike and that could coordinate a second strike or even post-war retaliation plan, then the problem of a compassionate human in the loop would be effectively removed. That seems a bit far-fetched; the real problem is the probability that a space-based weapons defense could be the anti-missile technology so feared by the architects of the now-defunct anti-ballistic missile treaty of the Cold War.

Space has extraordinary potential as a military staging base. It has been called the *ultimate high ground* because of its unflankable position atop the gravity well of Earth. In battle, the high ground provides a distinct advantage to whoever holds it. It is more difficult to fight uphill than down. Gravity gives extra impetus to arrows, bullets, or even rocks that are sent down at the opponent, a kinetic advantage that is difficult to overcome from below. Moreover, it gives an overview of the battlespace that is beneficial. As easy as it is to identify the high ground, the line-of-sight advantages of command, control, intelligence, and surveillance make possession of it a key force-multiplier.

In an example of the benefit of space as a high ground, it is estimated that a 6 m-long, 30 cm-diameter rod made of titanium or depleted uranium could simply be dropped or expelled from an orbiting satellite towards the Earth. With a heat-deflecting carbon-compound tip, the rod would impact the earth at speeds above Mach 10 with the explosive force of more than 11 kilotons of TNT – about the same strength as the Hiroshima bomb. If the speed could be boosted on launch or in flight, resulting in velocities in excess of 36,000 feet per second, the kinetic force (with no explosives or radioactive residue) would be over 100 kilotons of TNT and could penetrate up to 60 meters of hardened concrete. If dropped on a volcanic caldera or area of heavy surface thermal activity – Iceland or Yellowstone Park, for example – it could penetrate the earth's crust and initiate an eruption or earthquake (currently prohibited by international treaty). In another example, to show the relative strength of the earth's gravity well, if I were to walk out into my back yard one night and shoot a hunting rifle at the moon, I would only send the bullet a few miles into the air at best, after which it would fall back to the Earth. If I happened to be on the moon, however, and took a shot at the Earth, as long as my aim was true I could hit my target. (In theory – the bullet itself would burn up in the atmosphere long before it hit the ground – but it would get there.)

Kinetic weapons in space have lots of issues with targeting and timing, and would be extremely expensive. They also run a higher risk of causing unintended damage if used as anti-satellite weapons, potentially leaving a cloud of hazardous debris for space navigation, and so have gone out of favor. Current efforts to develop space weapons are more likely to be in directed energy; electrical lasers, concentrated microwaves, and plasma emitters. Once again, the high ground of space provides huge potential benefit. Have you ever wondered why a picture taken with a camera from a satellite in outer space shows exceptional detail, good enough to identify an individual person, but that same camera looking up from the earth could not properly determine the gross dimensions of a satellite in orbit, much less its technical characteristics? This is due to the so-called shower curtain effect. When moving from a more dense to a less dense medium, the distortion effect on light is at the front or beginning of its path, making the resulting distortion greater the farther one goes. When moving from a less dense medium, in this case a near vacuum in space, light travels relatively faithfully toward the target until the very end, when a denser medium is encountered (the air) and distortions start to multiply.

Since lasers are a form of light, they obey the same physical laws. Thus space-based lasers would have very little distortion as they pass through the vacuum of space, and a small beam could do significant damage to another satellite or projectile in flight without smashing it into an orbiting cloud of space junk. A similar beam could do significant damage from space to earth with minimal distortion as it passes through the atmosphere. But it would take a great deal of power, a *very* strong laser to do equivalent damage from earth to space. With a mix of both directed-energy and kinetic kill weapons based in orbit, the owning state (or international consortium of space-faring states) could offer both ultimate defensive protection and ultimate offensive violence to its security-desiring population. Day-to-day policing of high-value or fleeting targets could be accomplished with judicious use of directed energy. If the response was not sufficient, the opposing state might risk the terrible reprisal of titanium rods raining down from space – as powerful as nuclear weapons but without the widespread radiation or electromagnetic pulse (EMP) effects that can create widespread collateral damage and, so far at least, impossible to defend against.

Having such a heavy-handed deterrent has numerous problems, not the least of which is who should control such awesome power. So long as the wielder acts fairly and with restraint, particularly if it is controlled by a liberal nation-state or international consortium of like-minded states, it could be the basis for a long-lasting and stable peace. The peculiar thing is, even when individuals or states were never intending to violate the peace, or had planned to take a specific action for their own benefit, as soon as another demands they comply at the point of gun, they tend to resent and resist the pressure. And there is no way to ensure the wielder is benevolent. If power is corrupting, we should be extremely wary of anything approaching absolute power. When it comes to war, science will find as many ways to exploit and use such power as it does the means to counter it. And the beat goes on.

But fear of too much power is a human emotion. More nuanced problems were isolated in the nuclear era of the Cold War. President Eisenhower was a notorious fiscal conservative, and was convinced that the billions spent on a nuclear arsenal were wasted if a conventional force also had to be maintained. So he formulated a policy called Massive Retaliation. The US would draw down its conventional force and rely on its huge nuclear advantage. *Any* transgression, no matter how slight, *could* be met with a massive nuclear counter-strike. At first quite frightening, it soon became absurd. Really? Any transgression? The 1950s began to look like a child's game. What if I just do a little tiny bad thing, will I be nuked? No? How about something worse? Eventually the credibility of retaliation became completely discounted. The *only* thing the US would retaliate for with a massive nuclear strike was a nuclear strike against America. As posed in chapter 1, as most commentators believed no sane state would allow a conflict to escalate to a nuclear strike, the policy of

Massive Retaliation may have made the world safe for conventional war.

After Eisenhower, presidents demanded more deterrence options, and tasked scientific think-tanks to determine how to rebuild America's conventional forces. This is where lasers come into importance. The top space-faring nations, minus Japan, all have nuclear capabilities or could get nuclear capability soon. Their arsenals protect them from an unlikely nuclear strike from rational nuclear-armed competitors. But what of the *most likely* nuclear attacks? Do massive nuclear arsenals protect against these five basic categories: Rogue or pariah states led by dictators that may or may not be fully rational, especially if they are at the losing end of a future conflict (e.g. North Korea); accidental or unintended launches (there have been several close calls over the years, and ICBMs cannot be recalled; once launched they will strike their targets if not shot down); rogue military commanders who have control of nuclear weapons and choose to start a war (e.g. the Mad Boat Captain scenario that undergirds Tom Clancy's *Hunt for Red October* or General Ripper's actions in the movie *Dr Strangelove*); a well-funded terrorist gains control of a nuclear device (difficult, but plausible); and the most complicated and likely scenario, third-party nuclear war that draws the major powers in? How could the US,

China, or Russia deter an Israeli attack on Iran, or vice versa? What about a future conflict between India and Pakistan?

Space-based lasers have the advantage of requiring far less energy to achieve an effect on or above the surface of the earth than land-, sea-, or air-based lasers, by an order of magnitude (a minimum of ten times less power for an equivalent effect). This means a complement of 15 kW lasers, easily achieved and already commercially available at rapidly decreasing cost, could have deadly effects. Since laser power is 100 percent additive – three 15 kW lasers targeting the same point would have 45 kW of destructive power – a network of small lasers in space could have *scalable* burn-through capability in micro-seconds. Indeed, this is how medical science developed anti-cancer radiation therapy. When it was discovered in the 1930s that lasers could destroy biological tissue, they were applied to cancer cells. The problem was that a laser capable of destroying cancerous cells would also destroy non-cancerous ones, leaving a trail of destruction from entry to exit. The solution was to use several lasers, none of which had the capacity to destroy individual cells but all of which, when focused on a specific point from multiple directions, would converge on the cancerous cells without harming healthy surrounding tissue.

If deployed, such a networked space-weapons capability could assist in creating the *defense* necessary to give the world's population a sense of safety, at least from things that go boom in the night, and would have the *offensive* capability to punish attackers for their effort, adding to the *deterrent* effect of a defensive system. A small laser, or better yet, several dozens of small lasers networked in orbit around the world, would have the capacity to engage missiles or aircraft in flight *at the speed of light*. Unlike land- or air-based interceptors, which are limited to line-of-sight targeting missiles as they are incoming, a network of space-based interceptors would have global coverage, would not have to be based in foreign territory or given over to the control of others. As technologies get cheaper, lighter, and more efficient, more targets can be identified and engaged from space in near real time. Moreover, since the attack comes *from* space, there is no threat to the target state's sovereignty. No boots are deployed on the ground; no build-up of invasion forces on the border is required. The space-faring state(s) can police the world's borders, support blockades and embargoes; respond immediately to sudden or unexpected threats; dissuade or defend oppressed peoples; and engage high-value, fleeting targets as desired. All this depends on the power-wielding state deciding *not* to provide an international collective good and instead to simply transition to an empire, a historically difficult thing that has never lasted for long. If there is a chance that benevolence in such a situation can be expected, it may be with the promise of the democratic peace described in the next section of this chapter. Before that, however, the potential role of space science regarding the second and third broad motivations for war is offered.

Unlimited clean industry: Wealth from space

To the extent that satisfying all material wants might remove the second great motivation to war, only space development has the potential to spur a completely green neo-industrial age for all humanity, for only in space can the waste produced by industry and human habitation be permanently disposed of. This would require a major reformulation of current international space law, and I have argued elsewhere how this might be ordered, but the point is too far from the thesis of this short book to go into it here. If it *could* be negotiated – or simply proclaimed – space has the resources and potential to open up the world economy to an effectively infinite source of raw materials and clean energy.

The Sun provides such an abundance of energy that if all the light reaching the earth could be collected, just four and a half minutes of direct solar rays could power *all* of humanity's current energy needs for a year. Given such tremendous potential, if a large solar energy collector could be deployed in space – say 25 km in diameter (quite doable) – and the power collected transmitted to earth, via concentrated microwaves or perhaps some Tesla-inspired method (not yet doable, but quite conceivable) – free, practically unlimited clean energy could be available to all people everywhere. Fossil fuels would no longer have to be burned. Energy limits would no longer plague poverty-stricken areas. High-energy desalinization could be made available on a global scale. Deserts could bloom. Freed from energy limitations, entrepreneurs could develop as yet unimagined processes for enhancing the lives of people. Of course, vested interests in the highly profitable current energy provision system would have to be appeased, but this is a global good that would benefit all of humanity regardless of the disruptions to current social and economic stratification.

With unlimited free energy, new methods of accessing space could be fast-tracked. If a 100 percent reliable earth-to-space transportation could be devised, earth-based pollution could be quickly removed and permanently eliminated. Toxic and radioactive waste could simply be placed in barrels, lifted to orbit, and then pushed into a trajectory toward the Sun (a giant nuclear reactor) where it would be permanently eliminated. Possibilities for putting these barrels into space include magnetic rail guns that would propel the barrels rather than boost them on explosive rockets, and space elevators. The latter is essentially a massive cable anchored near the equator and extending out beyond geostationary orbit. Centripetal and centrifugal force would keep the cable suspended, and the static electricity collected by the cable as it passes through the atmosphere could power a lift system that would hoist the barrels from earth to a processing facility near the top. Given that the world spends many billions of dollars annually just to process and store such waste, with the storage compartments likely to break down and leak long before the toxicity is neutralized, waste removal would provide an instant monetary benefit.

With unlimited clean power and reliable space transport, raw materials from outer space could be collected and shipped to production facilities also in space. Whole comets and asteroids could be towed to these facilities and the minerals within them smelted and purified with all waste – including heat – permanently dissipated or collected and destroyed by solar immolation. Perfectly mixed and blended amalgams could be fashioned and transported to the earth's surface for use, limited only by the imagination of those who look to the stars for inspiration.

Of note, potential commercial development highlights why security needs precede interest or wealth desires when understood in their proper order, and why space weaponization is such a critical component of the future space economic boom. Pacifists – anti-war proponents as opposed to the more strident pacifists who practice anti-violence in all forms – lament the arming of the heavens, arguing that since war has not yet come to space it is humanity's duty to keep it martially pristine. Their argument is fallacious on two points: (1) war has long been going on in space, it is simply invisible from the earth, and (2) where there is no military to police the international commons, very little commerce can be sustained. In international waters today, for example, the US Navy patrols for pirates, aids ships in distress, provides free navigation and weather data, clears obstacles to transport, monitors for illicit commerce including human trafficking, and forces open internationally recognized trade routes illegally claimed by nation-states. The simple fact is that because the US military – and before it the British navy – is on the world's oceans, the likelihood that a container ship or a fishing vessel will get from one port to another or from its business back to home is far greater than if there were no military presence on the oceans. The US Air Force does much the same for international air space, and a future weaponization of space – if done in similar manner by a single space-faring state or, better yet, a consortium of such states – will *encourage* commercialization efforts by enhancing the likelihood that the fruits of exploration can be retained by those admixing their labor to it.

The best of humanity: Prestige from space

Even when all security and material needs are met, fatal arguments over honor and prestige do not fall away. Indeed, they become more petty, and because of their insignificance more hotly contested. Or so it seems. But when harnessed to a social good – and a clean, safe, abundant future for all humanity seems to meet the definition – it can enhance the broader productivity of the enterprise. Academics, for example, may nearly come to blows when splitting the thinnest of intellectual hairs, but they will rally to the defense of public education as a vital component of the nation's well-being. Scientists may be uncomfortable with the military's presence, but achievement in space science will bring tremendous fame and lavish state funding. They will participate.

Space has a long record of uniting humanity, even when the competition for firsts is at its peak. The best and brightest are selected to go forth, and the lure of exploration keen. When President Kennedy challenged the US to land a man on the moon and return him safely to earth, space science took off. Scientists and engineers flocked to NASA for minimal pay, and *they achieved the goal* (note how goal-oriented science – like war, when the ultimate goal is victory – was unable to continue robust space development once the "end state" of reaching the moon was attained). Competition with the Soviet Union was keen for several years, but there was little likelihood of going to war over space – the race was for prestige and the whole world was watching. What could be done to cause a future, similarly positive space race? Perhaps a Chinese Moon Base, forecast for the next decade, will set off a renewed emphasis on the stars. In the meantime, solving the problem of war has a glimmer of hope based on the investigation and analysis of social scientists.

The democratic peace

Perhaps the most promising social scientific discovery of the late twentieth century was the objective observation that in the modern era liberal democratic states have not gone to war *with each other* (Doyle 1983). Indeed, such states rarely even threaten violence against other democracies. More intriguing is that liberal democracies tend to go to war about as often as non-liberal states, and at least as violently – just not with each other. As the data continues to pile up, it is harder to ignore the possibility there is something about *mutual* democracy that acts as a brake on war.

When Immanuel Kant first argued that liberal republics would be far less likely to go to war than authoritarian or autocratic states, he didn't specify a relational qualifier. He was convinced that liberal republics would go to war now and again as some fighting appears necessary or unavoidable. In particular, he was concerned with expansionist and aggressive non-republics attempting to take land and resources from peaceful republics that would be obliged to resist. For the most part, however, Kant argued that the simple mechanism of including the people likely to do the actual fighting in the collective decision to go to war would make the state far less disposed to do so, and make it at least unlikely that republics would do so capriciously.

History has not been kind to Kant's theory. Democracies have waged horrendous wars that are no less violent than their non-inclusive counterparts. They have unleashed extraordinary forms of destruction and, under the proper circumstances, with far more enthusiasm than their non-democratic counterparts. Still, as the evidence continues to pile up, political scientists have searched for explanations of the phenomenon of mutual exclusivity. Two of these, outlined by Bruce Russett and John Oneal (2000), appear to be most persuasive, and probably work in tandem to strengthen the peaceful relationship between liberal democracies.

The first is a structural bias. Democracies require time to go to war. The public has to be made aware of differences with another state and propaganda or jingoistic media needs time to make a compelling case for war. A just cause must be persuasively offered. Non-democracies, on the other hand, require no such level of supporting preparation. Military forces that are on duty can simply be sent into combat without questioning from the population. In other words, non-democracies can conduct a *strategic surprise attack*, democracies cannot. Since democratic states know that a paralyzing attack can come at any time from an authoritarian state, they are wary and constantly on guard. But since they also know that a strategic attack by a fellow democratic state will be obvious in its preparation, time is available for a counter-preparation that may forestall or deter the anticipated war *and* for cooler heads to prevail. The merits and shortcomings of war are exposed; negotiations and alternative solutions can be publicly addressed and debated.

The second explanation is normative, and has to do with the manner in which politicians achieve power. In John Locke's consensual model of democratic government, the formation of the state is an act of political will, embodying the cardinal rule that the majority of the people, through their representative institutions, govern. The majority is neither arbitrary nor omnipotent, however; it is constrained by constitutional limits and recognized rights of individuals and minorities. The mechanism by which this works is a system of regular, free, and fair elections.

In liberal democratic states it is extremely rare that a chief executive will have risen to power without having one or more political setbacks in her or his career. Modern liberal democracies in this way rely on electoral losses as much as victories to maintain their robust health. The method for pulsing the public's preference, usually an election for office but also issue-specific referenda, means the side that loses *must submit* to the ruling authority of the state. The ruling authority is not omnipotent; the minority may be coerced only to the extent necessary to implement the basic constitutive agreement, but coercion cannot be used to silence or destroy the minority. And so the minority, the loser, *will* submit *because* it knows in the future it will have an opportunity to (fairly) compete again. If enough voters can be persuaded to its side, it will eventually prevail and be in the position of power.

This is how liberal democracy works. The majority has the right to act while the minority has the right to protest, and – critically – the *right to become the majority*. This governing form incorporates both the obligation to obey and the rights to criticize, protest, and oppose, according to Macridis and Brown:

> It allows the *force* of the state to be transformed into authority, deriving its legitimacy from the basic agreement. Individual dissent is expressed not in disobedience, but through organized opposition seeking to present alternative policies. Thus, opposition in the democratic scheme is harnessed to the total political system, which is strengthened, not weakened, by political dissent. (1972: 100)

Authoritarian leaders do not have such a background. Legitimacy is based on force, the capacity to coerce the population into obedience. Dissent is not tolerated, and where the dictator or hereditary leader is in power, dissenters have no reasonable hope of lawfully gaining power. Submission to the will of another is weakness, and weakness is the cancer of authoritarian rule.

So, liberal democratic leaders will come to power through consensus building, compromise, and tolerance of opposing views. Leaders of two states so constituted will be willing to accept that the other side may have legitimate grievance or claim, and that a political solution that works must be built on mutually acceptable terms. Authoritarian leaders will not accept consensus or compromise that is not decidedly in their favor, and even liberal democratic states will be wary of negotiating fairly with them – expecting defection and hard-line opposition. Thus wars between two liberal democracies are extremely unlikely whereas they may be quite likely if one or both adversaries are authoritarian.

Still, many social scientists are skeptical. Some argue that it is not democracy but capitalism that makes war unlikely. Since modern liberal democracies tend to favor free markets and the pursuit of wealth, war as a wealth-draining activity (as opposed to wealth-acquiring – war was perceived as potentially profitable prior to the heavy industrialization of the nineteenth century) is eschewed. Others argue that the core liberal democracies have always had a common external threat – monarchy, fascism, communism, now possibly fundamentalism – to unite them. When the majority of states are democratic, they predict, tensions will rise and war will become an increasingly viable option for them. These theorists don't argue *against* liberal democracy as a form of governing, they simply assert caution against *relying* on mutual democracy as a hedge against war.

And there does seem to be a caveat. In most cases, transitions to democracy appear to be extremely fragile, and such states tend to have violent episodes (Mansfield and Snyder 2002). But once established (and the rule of thumb is at least two successful national elections as long as there is at least one peaceful changeover of political power at the top), it is difficult to find a single case in all of history in which stable democracies have ever fought each other. So robust is the data that Spencer Weart (2000) has argued that *any* form of shared or republican government is mutually immune to war. If true, and the scientific data remains consistent, then the road to permanent peace seems clear. When all states are democratic, war will be relegated to the trash heap of history. Perhaps. It may turn out that someday a war will occur between two such constituted states, and that will be cheerless development. Even so, if war between democratic states is very, very difficult, and almost never happens, that is a vast improvement over the number and intensity of wars today. Science may not have created the democratic peace, but it observed, measured, and analyzed the phenomenon. Non-scientists will have to determine its human value.

"You must not blame us scientists for the use which war technicians have put our discoveries."

Lise Meitner

Science cannot abolish war because science has become war, and war become science. They are inextricable because they share the same absolute requirement – the relentless pursuit of more knowledge, more control, and more authority. Science provides the rational objectivity that war loses in practice through human emotion and passion, and is thus highly prized by warfighters. In peacetime science belongs to the world. In war, scientists belong to the state. If a solution to war can be found, it must come from *outside* science, probably from political agreement or subordination, or possibly from renewed religious or moral indignation. This is simply and unavoidably because the solution to war cannot be rational; and science cannot be irrational.

To the extent that war is a natural occurrence, an inevitable outcome of the interaction of individuals evolutionarily determined to be aggressive, the scientific solution must be to transform the essence of humanity. Options ranging from drug and psychological therapies to repetitive behavioral training all the way up to genetic manipulation and forced eugenics follow such logic. If war is not a natural human trait, then a level up from gross aggregations of individual characteristics are social and political scientists who see the tinderbox of war as the conditions established by *interacting* societies – in today's case nation-states. Whether looking for physical security or profitable gain, disputes of honor that may include social adjustments, reparation of grievances, adherence to common values or norms, states go to war for a myriad of context-driven causes, many of which are seen at the time and in retrospect as perfectly reasonable and just. The solution to war in this case is to remove the causes of imbalance in the system, those things that make adjustments (occasionally violent) to the status quo at least *appear* necessary. When imbalance is the cause of war, the solution is to create a situation in which change is measured and controlled and the balance is restored. If social dynamics are the problem, then social statics are the solution. Where social change is challenging the extant social order, create (an occasionally enforced) stability. Of course, wherever there is a prevailing status quo, there are those who are on top and those who are on the bottom, and to the latter the status quo will always be undesirable. They will seek change. The only long-term means to hold on to a permanent stability is to ensure that all people everywhere are equal – an impossible task, even for science. When change is constant, inequalities will proliferate.

Where war is perceived as the problem, as opposed to humanity, the cause is generally technologically determined. In this view, weapons cause violence, or at least spur normally passive individuals and states to more aggressive actions. The solution is to eliminate the weapons, and if this is impossible then to at least reduce the destructiveness and lethality of weapons. The questions then becomes: does decreasing lethality make it easier to go to war, or conversely, when one side has superior weapons does the prospect of an easy victory make war a desirable, perhaps efficient option?

So why not go for universal disarmament? Wouldn't that be a satisfactory response that science could certainly help with if not make happen on its own? Maybe. People and states that are prone to violence may be more likely to act out if they acquire the tools to do so. Once again we find an argument based on the essence of human nature. Is violence or cooperation the norm; war or peace the aberration? In a counterargument,

Hans Morgenthau stated that "Men do not fight because they have arms. They have arms because they deem it necessary to fight" (1967: 279). Then again, they deem it necessary to protect themselves from those who *do* have arms; weapons become vital for the defender, the side that desires peace *but not at any price*.

War as an evil to be excised from the body politic is not a universal point of view, however, and science has not yet been able to determine the rightness or wrongness of it. Indeed, a majority of commentators prior to the twentieth century considered war to be relatively beneficial, albeit brutal, purging the system of abnormalities and resolving disputes quickly and decidedly. In this view, social conflict is integrative. It establishes group identities, cohesion within and boundaries without. Some even suggest that war is a type of relief valve for society, expelling excess violence that builds up within groups and focusing it outward on others. The intensity or aggressiveness of a particular war is not wholly determined by these original causes, however. War ferocity is modified by religious or moral beliefs (shared or in conflict), by formal limits to combat, and long-standing relations that have led to past covenants. Hence war is viewed as a reasonable response to rectifying large inconsistencies when less radical adjustment efforts are likely to fall short of desired outcomes. It is neither good nor bad in itself, it is *instrumental*. When it is used to correct perceived wrongs, uphold the rights of an oppressed group, or settle disputes (if they are truly settled and not simply postponed), then the outcome may justify the method.

It may even be that the problem of war is not suitable for a proper scientific analysis. The classic configuration – if x, then y – isolating both the dependent (y) and independent (x) variables may not even be a useful form for investigating the problem from a medical, chemical, biological, or even physical scientific approach. *If* science, *then* war? *If* science, *then* not war? This shortcoming should not be seen as a condemnation of the scientific viewpoint that the world is knowable and that with enough systematic evidence and rigorous examination the mechanisms of human activity will be revealed – in this case, the causes of war will be known and a treatment plan devised that will both alleviate the symptoms of war and eliminate its roots. Rather, it should be seen as evidence that science, as incredibly positive as its impact has been in a multitude of human activities, cannot fully explain, adjudicate, or – critical for the study at hand – *solve* the problem of war *on its own* or *from within its own logic*.

We are left with one last conundrum. Science may not be able to *end* war, but can humanity move forward without science? Probably not. Science has revealed the knowledge and provided the technology that allows the earth to support a population far in excess of its natural carrying capacity, and there is no turning back without accepting the deaths of more than seven billion (and counting). With science, humanity can continue to expand, press the frontiers of space and, with its help, make life better for people to come. What is *better*, however, including a preference for war or peace in a given context, will be determined outside of science.

Ach, Johann, and Lüttenberg, Beata (eds.). *Nanobiotechnology, Nanomedicine and Human Enhancement* (London: Transaction, 2008).

"Active Denial Technology Fact Sheet," *US DOD Non-Lethal Weapons Program* (August 14, 2013). Online at http://jnlwp.defense.gov/PressRoom/FactSheets/ArticleView/tabid/4782/Article/577989/active-denial-technology-fact-sheet.aspx, accessed May 20, 2015.

Art, Robert. "To What Ends Military Power?" *International Security* 4 (Spring 1980): 3–35.

Bainton, Roland. *Christian Attitudes toward War and Peace: A Historical Survey and Critical Re-evaluation* (Eugene: Wipf and Stock, 2008).

Beason, Doug. *The E-Bomb: How America's New Directed Energy Weapons Will Change the Way Future Wars Will Be Fought* (Boston: Da Capo, 2005).

Bergin, Chris. "SpaceX Lifts the Lid on the Dragon V2 Crew Spacecraft," NASASpaceFlight.com (May 29, 2014). Online at http://www.nasaspaceflight.com/2014/05/spacex-lifts-the-lid-dragon-v2-crew-spacecraft/, accessed December 12, 2014.

Biagioli, Mario. "Patent Republic: Representing Inventions, Constructing Rights and Authors," *Social Research* 73(4) (Winter 2006): 1129–72.

Brinton, Crane. *Anatomy of a Revolution* (New York: Vintage, 1965).

Bibliography

Brodie, Bernard. *The Absolute Weapon: Atomic Power and World Order* (New York: Harcourt Brace, 1946).

Brodie, Bernard, and Brodie, Fawn. *From Crossbow to H-Bomb* (Bloomington: Indiana University Press, 1973).

Budiansky, Stephen. *Battle of Wits: The Complete Story of Codebreaking in World War II* (New York: Free Press, 2000).

Clarfield, Gerard, and Wiecek, William. *Nuclear America: Military and Civilian Nuclear Power in the United States, 1940–1980* (New York: Harper and Row, 1984).

Clausewitz, Carl von. *On War* (Princeton: Princeton University Press, 1976).

Cochrane, Charles. *Thucydides and the Science of History* (Oxford: Oxford University Press, 1929).

Cowley, Robert, and Parker, Geoffrey. *The Reader's Companion to Military History* (Boston: Houghton Mifflin, 1996).

Davenport, Thomas. *Big Data at Work: Dispelling the Myths, Uncovering the Opportunities* (Boston: Harvard Business School Press, 2014).

Davies, Wyre. "New Israeli Weapon Kicks up Stink," *BBC News*, Jerusalem, October 2, 2008. Online at http://news.bbc.co.uk/2/hi/middle_east/7646894.stm, accessed December 12, 2014.

Delbrück, Hans. *Medieval Warfare: History of the Art of War*, Volume III (Lincoln: University of Nebraska Press, 1990).

Dobyns, Henry. *Their Number Became Thinned* (Knoxville: University of Tennessee Press, 1983).

Doyle, Michael. "Kant, Liberal Legacies and Foreign Affairs," *Philosophy and Public Affairs* I and II (1983): 205–35; 323–53.

Drollette, Dan. "Blinding Them with Science: Is Development of a Banned Laser Weapon Continuing?" *Bulletin of the Atomic Scientists* (September 14, 2014). Online at http://thebulletin.org/blinding-them-science-development-banned-laser-weapon-continuing 7598, accessed December 12, 2014.

Bibliography

Enemark, Christian. *Armed Drones and the Ethics of War: Military Virtue in a Post-heroic Age* (New York: Routledge, 2013).

Enthoven, Alain, and Smith, K. Wayne. *How Much Is Enough?: Shaping the Defense Program, 1961–1969* (Santa Monica: RAND, 1971).

Fant, Kenne. *Alfred Nobel: A Biography* (Stockholm: Norstedts Forlag, 1991).

Freemantle, Michael. *The Chemist's War: 1914–1918* (London: Royal Society of Chemistry, 2014).

Friedman, Thomas L. *The Lexus and the Olive Tree* (New York: Anchor Books, 1999).

Giovannitti, Len, and Freed, Fred. *The Decision to Drop the Bomb* (New York: Coward McCann, 1965).

Gladstone-Millar, Lynne. *John Napier: Logarithm John* (Edinburgh: National Museum of Scotland, 2006).

Goody, Jack. "Industrial Food: Toward the Development of a World Cuisine," in Carole Counihan and Penny Van Esterik (eds.), *Food and Culture: A Reader*, 3rd edn (New York: Routledge, 2013), pp. 72–90.

Hambling, David. "Maximum Pain is Aim of New US Weapon," *New Scientist* (March 2, 2005). Online at http://www.newscientist.com/article/dn7077#.VJMtT_4cTxg, accessed December 12, 2014.

Hecht, Jeff. "Diode-Pumped Solid-State Lasers: Laser Dazzlers are Deployed," *Laser Focus World* (March 1, 2012). Online at http://www.laserfocusworld.com/articles/print/volume-48/issue-03/world-news/laser-dazzlers-are-deployed.html, accessed December 12, 2014.

Herken, Gregg. *Cardinal Choices: Presidential Science Advising from the Atomic Bomb to SDI* (Stanford: Stanford University Press, 2000).

Hölldobler, Bert, and Wilson, Edward O. *Journey to the Ants: A Story of Scientific Exploration* (Cambridge, MA: Belknap, 1994).

Bibliography

Howard, Michael. *War in European History* (New York: Oxford University Press, 1976).

Kaku, Michio. *Physics of the Future: How Science Will Shape Human Destiny and Our Daily Lives by the Year 2100* (New York: Anchor, 2011).

Kristensen, Hans, and Norris, Robert. "Global Nuclear Weapons Inventories, 1945–2013," *Bulletin of the Atomic Scientists* 69(5) (September/October 2013): 75–81.

Kuhn, Thomas. *The Structure of Scientific Revolutions*, 3rd edn (Chicago: University of Chicago Press, 1996).

Kurzweil, Ray. *The Singularity is Near: When Humans Transcend Biology* (New York: Viking, 2005).

Landes, David. *Revolution in Time: Clocks and the Making of the Modern World* (Boston: Belknap, 1983).

Lawrence, Paul. *Heisenberg and the Nazi Atomic Bomb Project: A Study in German Culture* (Oakland: University of California Press, 1998).

Lipson, Hod, and Kurman, Melba. *Fabricated: The New World of 3D Printing* (New York: Wiley, 2013).

Machiavelli, Niccolò. *The Prince and the Discourses* (New York: McGraw-Hill, 1960).

Macridis, Roy, and Brown, Bernard. *Comparative Politics: Notes and Readings* (Berkeley: University of California Press, 1972).

Mansfield, Edward, and Snyder, Jack. "Democratic Transitions, Institutional Strength, and War," *International Organization* 56(2) (Spring 2002): 297–337.

Massie, Robert. *Dreadnought: Britain, Germany, and the Coming of the Great War* (New York: Random House, 1991).

Mauroni, Albert. *Chemical and Biological Warfare: A Reference Handbook* (Santa Barbara: ABC Clio, 2007).

Mayer-Schönberger, Victor, and Cukier, Kenneth. *Big Data: A Revolution that Will Transform the Way We Live, Work, and Think* (Boston: Houghton-Mifflin, 2013).

Bibliography

McConnell, Malcolm. *Just Cause: The Real Story of America's High-Tech Invasion of Panama* (New York: St. Martin's Press, 1991).

Meitner, Lise. "Is the Atom Terror Exaggerated?" *Saturday Evening Post* (1946).

Moreno, Nathan. *Mind Wars: Brain Science and the Military in the 21st Century* (New York: Bellevue Literary Press, 2012).

Morgenthau, Hans. *Politics Among Nations: The Struggle for Power and Peace*, 4th edn (New York: Knopf, 1967).

Morley, Neville. *Thucydides and the Idea of History* (London: I.B. Tauris, 2013).

Morris, Christopher. *Academic Press Dictionary of Science and Technology* (Amsterdam: Elsevier, 1992).

Morris, Richard Knowles. *John P. Holland, 1841–1914: Inventor of the Modern Submarine* (Columbia: University of South Carolina Press, 1998).

Morton Szasz, Ferenc. *British Scientists and the Manhattan Project: The Los Alamos Years* (New York: St. Martin's Press, 1992).

Moskvitch, Katia. "Modern Meadow Aims to Print Raw Meat Using Bioprinter," *BBC News*, January 21, 2013. Online at http://www.bbc.com/news/technology-20972018, accessed on December 12, 2014.

Mumford, Lewis. *Technics and Civilization* (New York: Harcourt Brace, 1934).

Neufeld, Michael. *The Rocket and the Reich: Peenemünde and the Coming of the Ballistic Missile Era* (New York: Free Press, 1995).

Neufeld, Michael. *Von Braun: Dreamer of Space, Engineer of War* (New York: Random House, 2007).

North, Douglass. *Structure and Change in Economic History* (New York: Norton, 1982).

O'Connell, Robert. *Of Arms and Men: A History of War, Weapons, and Aggression* (New York: Oxford University Press, 1989).

Bibliography

Parker, Geoffrey. *The Military Revolution: Military Innovation and the Rise of the West, 1500–1800* (Cambridge: Cambridge University Press, 1988).

Paxman, Jeremy, and Harris, Robert. *A Higher Form of Killing: The Secret History of Chemical and Biological Warfare* (New York: Random House, 2011).

Perry, Walter, McInnis, Brian, Price, Carter, Smith, Susan, and Hollywood, John. *Predictive Policing: The Role of Crime Forecasting in Law Enforcement Operations* (Santa Monica: RAND, 2013).

Pogue, Forrest. *The Supreme Command* (Washington, DC: Office of the Chief of Military History, Dept. of the Army, 1954).

Rhodes, Richard. *The Making of the Atomic Bomb* (New York: Touchstone, 1986).

Roland, Alex. "Science and War," *Osiris* I (1985): 247–73.

Russett, Bruce, and Oneal, John. *Triangulating Peace: Democracy, Interdependence, and International Organizations* (New York: W.W. Norton, 2000).

Sample, Ian. "What is This Thing We Call Science? Here's One Definition...," *The Guardian*, March 3, 2009. Online at http://www.theguardian.com/science/blog/2009/mar/03/science-definition-council-francis-bacon, accessed on December 12, 2014.

Savulescu, Julian, ter Meulen, Ruud, and Kahane, Guy. *Enhancing Human Capacities* (Malden, MA: Blackwell, 2009).

Schultz, Colin. "Blame Sloppy Journalism for the Nobel Prize," Smithsonian.com, October 9, 2013. Online at http://www.smithsonianmag.com/smart-news/blame-sloppy-journalism-for-the-nobel-prizes-1172688/?no-ist, accessed on December 12, 2014.

Shanks, Peter. *Human Genetic Engineering: A Guide for Activists, Skeptics, and the Very Perplexed* (New York: Avalon, 2005).

Springer, Paul. *Military Robots and Drones: A Reference Handbook* (Santa Barbara: ABC Clio, 2013).

Bibliography

Stark, Rodney. *The Victory of Reason: How Christianity Led to Freedom, Capitalism, and Western Success* (New York: Random House, 2006).

Thomson, Thomas. *History of the Royal Society from its Institution to the End of the Eighteenth Century* (London: Robert Baldwin, 1812).

Thucydides. *History of the Peloponnesian War* (New York: Penguin, 1954).

Van Creveld, Martin. *The Transformation of War: The Most Radical Reinterpretation of Armed Conflict Since Clausewitz* (New York: Free Press, 1991).

Van der Kloot, William. "April 1915: Five Future Nobel Prize-Winners Inaugurate Weapons of Mass Destruction and the Academic-Industrial-Military Complex," *Notes and Records of the Royal Society of London* 58(2) (May 2004): 149–60.

Volkman, Ernest. *Science Goes to War: The Search for the Ultimate Weapon, from Greek Fire to Star Wars* (New York: Wiley, 2002).

Wawro, Geoffrey. *Warfare and Society in Europe, 1792–1914* (New York: Routledge, 2000).

Weart, Spencer. *Never at War: Why Democracies Will Not Fight One Another* (New Haven: Yale University Press, 2000).

Weinberger, Sharon. "Where's My Acoustic Bazooka?" *Wired* (April 1, 2008). Online at http://www.wired.com/2008/04/wheres-my-acous/, accessed December 12, 2014.

White, Robert. *Military Anecdotes and Other Sayings* (London: Carleton, 1995).

Wigglesworth, Jeffrey. *Science and Technology in Medieval European Life* (Westport: Greenwood, 2006).

Williams, Kathleen. "The Military's Role in Stimulating Science and Technology: The Turning Point," *FPRI Footnotes* 15(3) (May 2010). Online at http://www.fpri.org/articles/2010/05/militarys-role-stimulating-science-and-technology-turning-point, accessed December 12, 2014.

Winterbotham, F. W. *The Ultra Secret*, reissue edn (New York: Dell, 1975).

3-D printing, 111–16

Ach, Johann, 120
Active Denial System
    (ADS), 99–101
Advanced Step in Robot Mobility
    (ASIMO) 123–4
adventurism, 102
Agincourt, Battle of, 32–4
alchemy, as basis of modern
    science, 35–6
American Chemical Society, 64
American National Advisory
    Council for Astronautics
    (NACA), 64
anarchist movement, 59
anti-war, 50
Appert, Nicolas, 56
Archimedes, 27–9, 47
ARPANet, 130
Art, Robert, 143
Asculum, Battle of, 139
Autonomy in Weapons Systems
    Directive 3000.09, 127–8

Bacon, Roger, 30, 52–3
Bainbridge, Keneth, 3
Bainton, Roland, 53
balance of terror, 17, 145
Barnett, Thomas, 106

Beason, Doug, 102
Bergin, Chris, 114
Bhagavad Ghita, 3
Biaggioli, Mario, 45
Big Data, 130, 132–5
Big Dog robot, 126
Bohr, Niels, 70
Boko Haram, 18
Bonaparte, Napoleon, 27, 43
Brinton, Crane, 80
British Ordnance Department, 56
British Royal Society, 55, 62
Brodie, Bernard, 54, 85–6
Brodie, Fawn, 54
Brown, Bernard, 165
Budiansky, Stephen, 131
burning weapons, 99–101

Carnot, Lazare, 40–2
Catholic Church, 31
Charles V, 37
Charles VII, 33
Charles VIII, 34–5, 48
chemists' war, 60, 63
Chinese copy, 2
church bells and cannons, 31–2
Clancy, Tom, 152
Clarfield, Gerard, 1, 72
Clausewitz, Carl von, 15, 19, 77
Cochrane, Charles, 79

Cold War, and nuclear overkill, 17
Committee on Public Safety, 43
compellence, 13
  definition of, 144
Conant, James, 64
Cowley, Robert, 33
Cuban Missile Crisis, 13–14
Cukier, Kenneth, 132

Da Vinci, Leonardo, 53
Daksh robot, 126–7
Davenport, Thomas, 132
Davies, Wyre, 99
Declaration of Pillnitz, 41
Defense Department's Advanced
  Research Projects Agency
  (DARPA), 129
defense
  definition of, 143
  limitations of, 144
  from space, 154
Delbrück, Hans, 33, 37
democratic peace, 160–7
  arguments against, 165–6
  normative explanation for,
    163–5
  occurrence of war, 161–2
  robustness of, 166–7
  structural explanation for,
    162–3
designer pathogens, 117–18
deterrence, 13, 102, 105
  and credibility, 143–4
  definition of, 143
  space, 152–4
  versus defense, 143
Disney, Walt, 67
DNA manipulation, 119

Dobyns, Henry, 81
Doctors without Borders, 18
doomsday arsenals, 17
doomsday device, 146
Doyle, Michael, 161
Dr Strangelove, 152
Dragon V2, 114
Dreadnought, 57–8
Drexler, Eric, 121
Drollette, Dan, 95
Durand, Peter, 56

Ecole Polytechnique, 55–6
economic theory, classic, 5
Edison, Thomas, 16
Einstein, Albert, 70–3
Eisenhower, Dwight, 66, 128,
  151
electromagnetic pulse (EMP), 101,
  104, 150
  weapons, 101–2
Elizabeth I, 54
Enemark, Christian, 127
Engels, Friedrich, 8–9
ENIGMA cipher, 131
Enola Gay, 74
Enthoven, Alain, 17

Fant, Kenne, 51
first-strike, advantage of, 146
Formigny, Battle of, 33–4
forward-looking infrared (FLIR)
  system, 124
Francis I, 37
Franck, James, 61
Frederick the Great, 38
Freed, Fred, 3
Freemantle, Michael, 60
French Academy of Science, 42

Index

French and Napoleonic Wars, 39–42
Friedman, Thomas, 116

Galileo, 45
genetic splicing, 119
Giovannitti, Len, 3
Gladstone-Millar, Lynne, 52
Goody, Jack, 56
gray goo, 121
Gunpowder Revolution, 30–2

Haber, Fritz, 61
Hahn, Otto, 61, 70–1, 75–6
Hambling, David, 102
Harris, Robert, 118
Hecht, Jeff, 94
Heisenberg, Werner, 72, 74
Henry V, 32
Herken, Gregg, 73
Hertz, Gustav, 61
high-intensity directed acoustics (HIDA) weapon, 97
Hippocrates, 79
Hitler, Adolph, 69, 71–2
Holland, John, 58
Hölldobler, Bert, 84
Hollywood, John, 134
Howard, Michael, 41
humanitarian relief, 88
Hunt for Red October, 152

improvised explosive devices (IED), 101, 115
information war, 130–7
   and individual privacy, 133–6
   scarcity versus abundance, 131–3

intercontinental ballistic missile (ICBM), 67
   space defense from, 153
International Space Station (ISS), 114
Islamic State, 18

Kahane, Guy, 119
Kaiser Wilhelm Institute, 61
Kant, Immanuel, 161
Kennedy, John, 13, 138
Khrushchev, Nicolai, 14
knowledge, from inspiration, 27
   as power 27–8
Kristensen, Hans, 14
Kuhn, Thomas, 22
Kurman, Melba, 112
Kurzweil, Ray, 135

Landes, David, 32
laser weapons, 94–5, 96
Lawrence, Paul, 72
Leibniz, Gottfried, 16
Leveé en masse, 39, 42–3
liberal democracies and war, 163–4
   mechanism for peace, 165
Lipson, Hod, 112
Locke, John, 163
Lüttenberg, Beata, 120

Machiavelli, Niccoló, 11–12, 142
Macridis, Roy, 165
Mad Boat Captain scenario, 152
Mandela, Nelson, 138
Manhattan Project, 1–3, 6, 65, 69, 73, 75
Mansfield, Edward, 166
Marne, Battle of, 60

Index

Marx, Karl, 8
Massive Retaliation policy,
   151–2
Mauroni, Albert, 118
Mayer-Schönberger, Victor, 132
McConnell, Malcolm, 98
McInnis, Brian, 134
Meitner, Lise, 70–1, 75, 168
micro-war, 116–22
microwave weapon, 99–101, 103
Military-Industrial Complex, 69
Monte San Giovanni, siege of,
   35
Moreno, Nathan, 91
Morgenthau, Hans, 171
Morley, Neville, 79
Morris, Christopher, 7, 58
Morton Szasz, Ferenc, 1
Mosca, Gaetano, 26
Moskvitch, Katia, 112
Multi-function Agile Remote
   Control Robot
   (MARCbot), 126
Mumford, Lewis, 53
Mutually Assured Destruction, 14,
   20–1

nanobiotechnology, 120
nanobots, 121–2
Napier, John, 54–5
National Defense Research Council
   (NDRC), 68
nationalism, 42
nation-building, 88
Nernst, Walter, 60
Neufeld, Michael, 65–6
Newton, Isaac, 16, 44, 47
Nobel, Alfred, 51–62
   the Merchant of Death, 52

Nobel, Ludwig, 52
Nobel Prize, 52, 60–1
non-lethal anti-materiel weapons,
   91–2
   biological weapons, definition
      of, 91
   chemical weapons, definition of,
      91
   directed energy weapons,
      definition of, 91–2
   physical weapons, definition of,
      91
non-lethal anti-personnel weapons,
   92–5
   biological weapons, definition
      of, 93
   chemical weapons, definition of,
      92–3
   directed energy weapons,
      definition of, 93
   physical weapons, definition of,
      92
non-lethal weapons, 87–91,
   110–11
   definition of, 90
Noriega, Manuel, 98
normal science, 22–3
Norris, Robert, 14
nuclear arsenals, US and USSR,
   14
nuclear war, 85–7

Obama, Barack, 123
O'Connell, Robert, 29
offense, definition of, 144
Operations Other Than War
   (OOTW), 89
Oppenheimer, J. Robert, 1–3,
   73–5

Index

Packbot, 127
Parker, Geoffrey, 33, 38
patent laws, 44–7
  and US Constitution, 45–6
  and Venice, 45
Pavia, Battle of, 37–8
pax atomica, 17
Paxman, Jeremy, 118
peace-keeping, 88
Perry, Walter, 134
Planck, Max, 71
Pogue, Forrest, 66
police power versus military power,
  88–90, 102–4
Predictive Policing (PredPol), 134
preventive medicine, 80
Price, Carter, 134
Project RAND, 129
pulsed electromagnetic projector
  (PEP), 102
Pyrrhus, 139
  and Pyrrhic victory, 139

Ramsay, William, 62
Ratblat, Joseph, 74
Reagan, Ronald, 12
Red Crescent, 18
Red Cross, 18
remotely-controlled robot vehicle
  (RCRV), 108
remotely-piloted vehicle (RPV),
  109, 124
reversible-effect weapons, 87–8,
  110
Rhodes, Richard, 2–3, 70, 74
Rice, Andrew, 102
robot war, 123–30
Roland, Alex, 63–4
Roosevelt, Franklin, 68, 73

Sample, Ian, 7
Savulescu, Julian, 119
Schopenhauer, Arthur, 50
Schultz, Colin, 52
science
  and alchemy, 30
  as critical intervening variable,
    141–2
  definition of, 7
  goal of, 139–40
  intricately linked with war, 28
  limitations of, 22–3
  logic of, 138–9
  as modern faith, 19
  and modern liberalism, 46–9
  purpose of, 8
  and secrecy, 43–5
  and solution to war, 12,
    168–72
  as a way of knowing, 7–8
scientific method, 7–8, 78–9
  and moral values, 21–2
second strike, 145–6
Seven Years War, 38
Shanks, Peter, 119
sight weapons, 94–6
skunk weapon, 99
smell weapons, 98–9
Smith, K. Wayne, 17
Smith, Susan, 134
Snyder, Jack, 166
sonic bullet, 97
sound Weapons, 97–8
space elevator, 157
space power
  and neo-industrial age, 155
  prestige from, 159–60
  security benefits of, 142–3
  and toxic waste removal, 156–7

Index

space weapons, 146
  advantage of, 148
  and directed energy, 148–50
  kinetic bombardment, 147–8
  lasers, 148–50, 152–4
  policing space industry, 158–9
  and shower curtain effect, 149
  as ultimate high ground, 147
space-based solar power, 155–6
Space-X Corporation, 113
Spanish Armada, 39, 54
Springer, Paul, 124
Star Wars, 12
Stark, Rodney, 32
stealth technology, 92
Strassman, Fritz, 70–1, 75
Strategic Defense Initiative, 12
strategic surprise attack, 162
strobe weapons, 95–6
super-empowered individual, 116
Szilárd, Leó, 72–4

Teller, Edward, 75
ter Muellen, Ruud, 119
terrorists, 5
Tesla, Nikola, 16
Thomson, Thomas, 55
Thucydides, 11–12, 79–80, 142
Titterton, Ernest, 2
Trinity test, 1–3, 5–6, 74
Truman, Harry, 75

UK Science Council, 7
ULTRA, 131
UN Protocol on Blinding Laser
  Weapons (1995), 96
universal disarmament, 170–1
Utopian versus Realist goals, 20–1

V-2 Rocket, 66
Van Creveld, Martin, 32
Van der Kloot, William, 61–2
vengeance weapons, 65
Versailles Treaty, 65
Volkman, Ernest, 26, 32, 35, 53, 55
von Braun, Wernher, 65, 67

war
  and aggression, 81–2, 84
  causes of, 11–18
  character of, 6, 111
  definition of, 8–9
  as a dismal science, 26
  and ethos, 23–4
  as honorable, 26
  and legitimacy, 9
  as limiting, 10–11
  logic of, 138–9
  and oil, 15–16
  as political action, 82–3
  as political solution, 9
  preventing, 10–11
  as a problem of diagnosis, 10–11
  and problems for science, 168–72
  purpose of, 10
  and quantitative analysis, 86
  as rectifying 171–2
  and science, intricately linked, 28
  as scientifically-calculated, 41
  suitability for scientific analysis, 172
  violent, deadly, and destructive, 110
Wawro, Geoffrey, 39

Index

Weart, Spencer, 166
Weinberger, Sharon, 97
White, Robert, 139
Wiecek, William, 1, 72
Wigglesworth, Jeffrey, 30
Williams, Kathleen, 68

Wilson, E. O., 84
Winterbotham, F. W., 131
World War I, 4–5, 49, 58–64
World War II, 10

YouTube, 18, 113